Tapestries of Life

Tapestries of Life

*Uncovering the Lifesaving Secrets
of the Natural World*

Anne Sverdrup-Thygeson

Translation by Lucy Moffatt

Mudlark
HarperCollins*Publishers*
1 London Bridge Street
London SE1 9GF

www.harpercollins.co.uk

HarperCollins*Publishers*
1st Floor, Watermarque Building, Ringsend Road
Dublin 4, Ireland

First published by Kagge Forlag AS 2020

This UK edition published by Mudlark 2021

1 3 5 7 9 10 8 6 4 2

Text © Anne Sverdrup-Thygeson 2021
Translation © Lucy Moffatt 2021

Illustrations © Tomislav Tomić/Arena 2021

Anne Sverdrup-Thygeson asserts the moral right
to be identified as the author of this work

A catalogue record of this book is
available from the British Library

HB ISBN 978-0-00-840273-0
TPB ISBN 978-0-00-840274-7

Printed and bound in Great Britain by
CPI Group (UK) Ltd, Croydon

MIX
Paper from
responsible sources
FSC
www.fsc.org
FSC™ C007454

This book is produced from independently certified FSC™ paper
to ensure responsible forest management.

For more information visit: www.harpercollins.co.uk/green

What I discovered was that everything which meant most to me as a naturalist was being threatened, and that nothing I could do would be more important.

Contents

Preface

I was the kind of kid who asked questions about everything. All the time. A chatterbox, eternally inquisitive and doubtless obnoxiously precocious at times. In primary school I had an autograph book. A hideous bright green with a big floral pattern on its soft cover, typical 1970s. In among the usual 'Roses are red, violets are blue' and 'By hook or by crook, I'll be first in this book', inscribed in felt tip in my school pals' neat joined-up writing, there was a double page entirely taken up by my brother. He'd written a poem for me, which started like this: 'You've asked me time and time again, most likely googolplex ...' and continued with a list of all the stuff I used to ask him about.

Googolplex isn't just an enormously big number – 10 to the power of one followed by a hundred zeros (more than all the atoms in the universe); there is also something magical about the word itself, a bit like a spell. And I collected lovely words as a child: words that had a marvellous way of rolling and tumbling around in my mouth when I said them, like 'onomatopoeic', or the ones that hop-scotched from my uvula and across my tongue to land on its tip, like 'trigonometric point'. My grandfather introduced me to more fascinating words, including the Latin names of plants – like *Tussilago farfara* for coltsfoot. In summer time, high up on the Norwegian mountain plateau of Golsfjellet, he'd show me quartz crystals, the place where purple saxifrage grew and how the golden plover sang. My grandfather lived to 102 and I still think of him every summer when I hear the golden plover's mournful cry, high up above the treeline. Back home in Oslo, he would sit in the wing chair in the corner of our living room reading *The Cormorants of Utrøst* aloud from the two-volume fairy-tale collection. As I grew, so too did the scope of our conversations; he told me about Loki and the mistletoe, Jason and the Golden Fleece, transatlantic crossings to America by ship in the 1930s, the two world wars ...

My family had a cabin on a little island out on a lake in the forest where I spent holidays and countless weekends. It was a two-room log cabin with no electricity or running water but it was right up close to nature. Summers were filled with the tarry aroma of sun-warmed cabin walls, posters of mushroom species in the outhouse, perch caught in fish traps, bristling with tiny bones, wild strawberries from the cabin roof, wood-chopping and the tedium of enforced lingonberry-picking trips that never seemed to end. I read my way through adventure books for boys, their covers slightly mildewed from the damp of the boathouse where they were stored.

Since the cabin was a long way from the nearest community – indeed, a long way from the nearest neighbour – the skies were wonderfully starry in winter. As a teenager, I once made a bed of spruce branches out on the ice for me and my friend Nina, then dug out some wartime-era sleeping bags from the shed and organised an outdoor sleepover so we could look at the stars. Forty years later, my clearest memory of that night is not the Milky Way but the strange, crackling dry object my bare feet encountered at the bottom of my sleeping bag – an object that proved, on further investigation by the light of my torch, to be a mouse nest full of dead and mummified pups ...

Sometimes people ask why I'm so keen on writing about insects and other seemingly insignificant creatures with PR problems; whether I'm one of those people who collected bugs as a child. I'm not. But I was lucky enough to grow up in a family that spent a lot of time outdoors as a matter of course and was interested in the stories and language that describe the relationship between us and nature, past and present. And a family that allowed me to be curious and tried to answer my never-ending questions about how everything fitted together.

Curiosity and a capacity for wonderment are also important to me as a scientist. Since I am a professor of conservation biology, the science that studies threats to biological diversity and ways of dealing with them, I've wondered a lot about how to get people to appreciate the natural world around us so that we'll all want to take care of it too. This book is my attempt at an answer: I want to show you all the things that the wonderful natural world does so that you can see what's at stake. And I want to point out the paradox in our creative relationship with nature: we have made use of it, but our ability to exploit the benefits of nature also risks undermining the very foundations of our own existence.

Introduction

A Hornless Rhino

Some years back, I attended a science conference in Dublin. In between lectures about pollination and malaria mosquitoes, I took time out to visit the city's natural history museum, the National Museum of Ireland. I love museums and this one was especially interesting: cases containing insects collected by Darwin himself; the skeleton of a giant red deer, its antlers broader than I am tall – a melancholy monument to an extinct species; an exhibition containing several hundred astonishingly delicate glass models of marine invertebrates, created in the 1800s by German glass artists, the Blaschkas.

The glass figures were made as teaching aids, because it was difficult to display these sea creatures in any other way – sea anemones and soft corals tended to end up as shapeless, colourless lumps in the bottom of a jar of formalin. Several thousand of these elegant artworks were sold to museums, universities and schools throughout the world, and those that have survived are quite a sight.

But what really sent shivers down my spine was a stuffed rhinoceros; a rhinoceros without horns. In the place where the horns should have been, there were just two holes in its dark skin, which allowed me to see straight through into the rough, yellowish-white cotton canvas beneath. Beside the mutilated animal was a plaque. The museum apologised for the appearance of the rhinoceros and explained that the horns had been removed owing to the risk of theft. This is because of the utterly false belief that powdered rhino horn has medicinal properties, a belief that remains widespread despite the fact that the horn is made of keratin, the same substance that forms your fingernails. The horns are traded illegally all over the world and the people involved in this illegal market are totally unscrupulous: poaching, raids on museum exhibitions and large-scale smuggling are all commonplace. And neither buyers nor sellers appear to care that the product comes from a species that is in the process of vanishing off the face of the Earth – for good.

Perhaps this example illustrates an extreme version of an underlying attitude to nature and species diversity that I believe is shared, often unconsciously, by many people: to the extent that they think about nature at all, they view it as a kind of remote, impregnable bank of resources. A place that is set apart from us humans, with our comfortable existences and everyday lives – a service centre, where we can access limitless resources and which we expect to provide unrestricted services whenever we want it to, but which is otherwise of little concern to us.

That is not how it is. You and I are much more tightly woven into the wickerwork of nature than you think. Nature, with its countless tiny, barely visible organisms, is what is holding you up, holding your life together – even in our modern, increasingly urban existence. And the planet still abounds with species. To date, we have named around 1.5 million of them (excluding microorganisms), but we know there are many more – close to 10 million species. And we humans are just one of them.

The majority of the planet's species are nowhere near the size of a rhino and you've never met them because they are small and live their lives hidden away from us humans, down in the muggy warmth of the soil, among the fibres of dead and decaying wood, swimming in the saltwater of the sea. Nonetheless, this diversity of anonymous organisms is what you have to thank for being alive. They have been turning up for duty since long before the first human rose up onto two legs – and we have been taking their contributions for granted ever since.

On the Shoulders of Nature:
Ecosystem Services

In recent years, scientists have started to use a concept that aims to reveal how nature, in all its teeming diversity of organisms, contributes to our welfare. It goes by various names: ecosystem services, nature's goods and services or NCP: 'Nature's Contribution to People'. Regardless of which term you use, the principle is the same: it refers to living nature's direct and indirect contributions to human existence and welfare. All the benefits that nature offers.

And just as there are different terms for it, there are also different ways of categorising the benefits of nature. One much-used method is to classify them into *provisioning services*, *regulating services* and *cultural services* (note that if we opt to talk about nature in this way, from the perspective of utility to us, there will also be *disservices* – the dissemination of pollen as a problem for people with allergies, for example).

To describe these categories in a slightly more easily understandable way, let's put it like this: provisioning services are about nature as an old-style general store and apothecary, a place where we can pick up all kinds of products that we need: drinks, like clean water; food and fibre; fuels and active ingredients for industry; and raw materials for new medicine.

Regulating services are about nature as a trusty care-taker that sees to the clearing up and recycling – to ensure that water, soil and snow stay where they are supposed to and that temperatures don't go off the scale. Some of these functions are so fundamental to life on Earth that we could think of them as central strands in the very fabric of life, like the natural cycles of water and nutrients that endlessly repeat.

Cultural services are about nature as a source of knowledge, beauty, identity and experiences. We can learn about the past by studying nature's archives, in bogs or tree rings. We can draw inspiration from nature and come up with ideas for new ways of solving problems. For many, nature is also like a cathedral, a starting point for inspiration, reflection and awe, whether or not we assign any religious significance to it.

Life in an Apple Peel

In a sense, life and species diversity on Earth are robust. After all, life has existed here for billions of years. But the biosphere, that thin layer around the Earth within which life exists, does not extend especially far. Picture an apple and imagine the thickness of the skin relative to the size of the fruit. In proportional terms, that apple skin is actually thicker than the layer of living life on our planet. The height difference between the eternal darkness of the Mariana Trench at the bottom of the Pacific Ocean, the deepest place on Earth, and the snow-clad peak of Mount Everest is no more than 20km. All of our civilisation, from pyramids and cave paintings to toasters and the UN Assembly, relies 100 per cent on this thin stratum, where there is room for life.

Today, we find plastic bags in the Mariana Trench and tonnes of rubbish strewn across the faces of Mount Everest. We are many, we consume a lot and we are spreading ourselves out unashamedly. Three-quarters of the planet's land surface has been significantly altered by human action and we have filled this altered area with ourselves and our domestic animals. If we were to weigh all the mammals on Earth at this precise moment, our livestock – cattle, pigs, poultry and others – would account for more than two-thirds of this biomass. We humans alone account for almost a third. This means that wild animals of all sizes, from elephants to shrews, constitute only 4 per cent of the total weight of all living mammals.

I stood for a long time gazing at the mutilated rhinoceros in the Dublin museum, the one without horns. Anger and grief mingled to form a knot in my stomach.

At the bottom of the plaque there was one more sentence: the real horns would soon be replaced with plastic replicas. But perhaps the rhinoceros should stay the way it was, as a thought-provoking symbol of our inability to relate to facts and use our native intelligence, or to take other species into account, not even when they are on the brink of extinction. As a reminder that we humans must change our way of life if we wish to secure the foundations of our own existence.

We are just one species among 10 million. At the same time, though, we are unique in our ability to interact with one another in a way that enables us to make an impact on the entire planet and every other species. We have also, uniquely, evolved the capacity to evaluate our actions logically and morally from a greater perspective. With this insight comes great responsibility, and it is time for us to shoulder that responsibility – because nature is all we have, and all we are.

CHAPTER 1

Water of Life

Water is fundamental to life as we know it. We have yet to find a single organism that does not depend on it. This is partly because of water's versatility. It easily dissolves other substances and transports them around an organism, it is crucial to ensuring that proteins behave themselves properly in that same organism, and it exists in all three phases of nature (solid ice, liquid water, gaseous water vapour). In addition, water expands when it freezes, so it ends up floating on the surface of lakes and seas instead of laying itself like an ice-cold compress on the bottom.

You yourself are two-thirds water and have to top yourself up with several litres every day to ensure the proper functioning of your body. What's more, you use water for washing and other tasks. All told, per capita water consumption in the United Kingdom is 141 litres – equivalent to just under a bathtub – a day.

Some 71 per cent of the Earth's surface is covered in water, yet drinking water is a scarce benefit: only 3 per cent of the water on Earth is freshwater, and almost all of that is tied up beneath the penguins' feet in the Antarctic. This means that just 1 per cent of Earth's water is available as drinking water.

To be safe for human consumption, this water has to be clean, yet that is far from a given. On a global basis, one in three people lack access to clean, safe drinking water. Water is continuously cleaned and filtered every second as it flows, runs, babbles, drips or seeps through nature in the eternal round of the water cycle. Whole hosts of species – bacteria, fungi, plants, small creatures like mosquitoes and mussels – show up for work in nature's water purification system and try to keep pace with pollution, erosion, climate change and other factors, producing the clean water in our taps or wells. But when we degrade these natural systems, they're not able to keep up. This chapter is all about the task of water purification as well as the species whose hidden, silent work ensures that we get clean drinking water.

New York City: The Champagne of Drinking Water

I've been to New York City a couple of times. Every time, I'm drawn to Central Park, that lush green oasis in a landscape otherwise so thoroughly shaped by humanity. The good thing about travelling there from Europe is that you wake at the crack of dawn and have time for a run before the day begins.

We're not exactly talking about a wilderness here. Even the lawns have opening hours. They are closed after dusk and don't re-open until nine o'clock, as a sign informs me. But despite the closed lawns and early hour, there's a steady trickle of runners along the outer, tarmacked jogging trail. I want to find a running route that is softer underfoot, with fewer people, so I turn off onto a small walkway into the area known as The Rambles – a less well-coiffed part of the park. At an intersection on the path, a young girl with a ponytail bends over a drinking fountain. I stop and wait until she's drunk her fill because I just have to taste this water. New York City is famed for its fantastic drinking water, among the best in the entire US. And this is because NYC is one of just five major US cities that supply tap water straight from nature, without channelling it through a filtration plant.

The city's drinking water system is, in fact, the largest unfiltered water supply system in the world: around 4 trillion litres of water are delivered daily to the city's roughly 9 million inhabitants. Cities are thirsty things. There is washing and showering and drinking to be done. With all its skyscrapers and tarmacked avenues, its system of subterranean pipes and its hi-tech devices, the city is like a concentrated human-made terminus for a vast water collection area. This watershed spreads out in a fan of forests and mountains and some agricultural land – in all, nearly a thousand times larger than the surface area of Manhattan. Rain and meltwater from thawing snow seep through the vegetation and soil before reaching small streams, passing into rivers and ultimately finding their way into lakes and reservoirs. From there, they enter a system of tunnels and aqueducts, some of it dating back to the 1800s, before ending up in the city and my Central Park drinking fountain.

In the 1990s, the US federal authorities passed new legislation that set stricter requirements for the purification of drinking water, at a time when development and more intensive agriculture in the watershed were a growing problem for water quality. A treatment facility for New York's drinking water would cost something in the region of $4 billion, and involve an annual operating budget of around $200 million. The expenditure would lead to a doubling in the city's water bills. But there was an alternative ...

A unique collaboration came about between New York City and the municipalities and landowners in the watershed, driven by the newly recruited Commissioner of the New York City Department of Environmental Protection and Director of the New York City Water and Sewer System.

Large forest and wetland areas in the watershed would be left undeveloped. The agricultural areas already in use would be cultivated using environmentally friendly methods. Through a series of negotiations and agreements, arrangements were made for NYC to compensate for the additional expenditure this would entail. The city also bought up considerable amounts of land in the watershed, so the water quality was assured as forests and vegetation filtered and cleaned the water en route from precipitation to kitchen tap. In combination, these measures made the treatment facility unnecessary because natural processes and hordes of volunteers – bacteria, fungi, invertebrates and other tiny species in the ecosystem – did the job for free. At the same time, the land was secured as a habitat for biodiversity, and a resource for recreation and outdoor activities. Even so, total expenditure amounted to a fraction of the cost of a treatment facility.

In fairness, it must be said that this solution is neither easy nor exclusively positive. One issue is that the agreements are challenging to negotiate and require constant follow-up. What's more, larger beaver and deer populations create problems because these animals can be hosts for organisms that cause stomach upsets and diarrhoea in humans. It is debatable whether such microorganisms will be blocked by the chlorine and UV radiation that the water must still undergo, despite the exemption from filtration.

The bodies responsible for the city's potable water are therefore discussing whether these populations need to be controlled. So even this natural solution involves intervention from humans – adjusting nature to suit our needs.

In 2017 the system was put to the test. NYC had to renew its exemption from the strict federal regulations requiring the filtration of potable water. A great deal was at stake. The estimated cost of building a treatment facility now stood at more than $10 billion, with high operating expenses on top of that. This time too, though, the natural solution won out. In exchange for promises that a further $1 billion would be invested in improved septic systems, additional land purchases and support for environmentally friendly operating methods in the watershed, the city was authorised to continue letting nature do what it has always done: clean the water and make it pure enough for us humans to drink.

As I wait my turn at the Central Park drinking fountain, I jog on the spot and think about the hidden role nature plays, even here in the heart of a megacity like New York. I wonder whether the ponytailed girl is sending a friendly thought to the watershed in the Catskill Mountains now that she has finally drunk her fill. Possible, but unlikely. She just says hi in that cheery way Americans have, and runs onward, thirst quenched. At last, it's my turn to taste the famed 'champagne of drinking water', as New Yorkers like to call it.

Freshwater Pearl Mussels – Caretakers of the Water System

Few Norwegians have been given new names and identities by the authorities to be left in peace by people who pose a threat to their lives. Even fewer species have received the same treatment. The freshwater pearl mussel – previously known in Norwegian as the river pearl mussel, now simply the river mussel – may well be the only species that has been given a new name for this reason. The species owes its previous moniker to the fact that it sometimes contains a pearl. True, you have to open (and thereby kill) a thousand mussels to find a single pearl, and only one in a thousand are good quality – but that was no obstacle to centuries of intensive pearl fishing in both Europe and North America.

The freshwater pearl mussel is a freshwater mollusc reminiscent of the saltwater mussel. It is brown and lives half buried, standing on end in the riverbed. Just as filtration through plants, trees and organisms in the soil is the land-based part of nature's purification plant, this mussel species is part of the water-based filtration system. A single specimen can clean 40–50 litres of water every 24 hours. A riverbed covered in thousands of mussels that capture all kinds of passing particles and detritus really speeds up the water purification process. Sadly, the species is threatened, both nationally and globally. Norway is responsible for what we estimate to be a third of all the freshwater pearl mussels remaining in the world. Some of them have been alive since the US declared its independence more than 200 years ago.

In the Middle Ages, the Church, European royals and the Tsar's family in Russia were the ones most inclined to indulge in pearl embroidery and pearl ornaments. Apparently, some priests' vestments from European monasteries are sewn with more than 10,000 pearls. And if you ever see the painting of Queen Elizabeth I known as *The Armada Portrait* – the background depicts the English fleet's crushing victory over Spain in 1588 – just try counting the pearls on Elizabeth's dress, hair ornament and adornments. That'll give you a good idea of the Tudors' extravagance with pearls. It seems that these specimens came from the rivers of northern England and Scotland, although the Queen apparently had to resort to artificial pearls, too: glass beads dipped in glue and fish scales.

Around 50 years later, the bailiff of what is now Agder County in southern Norway sent a handful of pearls to the Dano-Norwegian King Christian IV and asked for instructions on the purchase of pearls. The King replied in a letter dated 27 June 1637, expressed in the elegant phrasings that were the officialese of the day. If you have trouble understanding modern-day communications from official channels, try this for size: 'Concerning the pearls which thou hast humbly dispatched to us, which the peasants who are subject to thee sell to strangers on acquiring them, and about which thou art desirous of knowing our most gracious wish: in such cases, we graciously wish that, whensoever the peasants might find such pearls, thou shalt pay money for them and see to it that no strangers acquire these same pearls, which thou shalt then dispatch to us.' Or put simply: 'Buy all the pearls and send them to me.'

The command was only partially effective as the peasants generally got a better price elsewhere. In any case, the pearls proved to be a short-lived source of income: by as early as the 1700s, the rivers were empty of these mussels and for a long time there was no profit to be had from harvesting them. But here's a fun fact for you: Norway's eight-pointed Crown Prince's coronet, made in 1846 and displayed for your admiration at the Archbishop's Palace in Trondheim, has a Norwegian freshwater mussel pearl on the tip of each point. The crown was meant to be worn by the Crown Prince at the coronation of the Swedish-Norwegian King Oscar and Queen Josephine in Nidaros Cathedral in 1847. However, as Josephine was Catholic, the Bishop refused to crown her and the ceremony was called off – so those eight mussel pearls never got their chance to add the finishing touch to the Crown Prince's stately garb.

Nowadays, pearl-gathering isn't the issue threatening the remaining mussels but old age. Freshwater pearl mussels can live to a great age, up to 300 years. That means Norway's oldest specimens were born in the 1700s, not so long after King Christian IV – of the elegant letter – ruled the land. By the time these mussels have made it to maturity, they can cope with a bit of rough treatment. But mustering new recruits is tougher because freshwater pearl mussels have a peculiar childhood, where their first obstacle is getting into 'kindergarten'. And their kindergarten is located in the jaw of a passing salmon. The mussel larva's survival depends on being able to spend a good year attached to the gills of a salmon or trout, before releasing its grip and digging down into the sediment, where it remains for several years.

This early phase of the mussel life is what is failing today. Pollution and erosion from agriculture, logging and other land use mean that there is too much silt or too many nutrients in the river and, as a result, too little oxygen. Consequently, the young mussels buried in the riverbed are suffocated. A decline in host fishes has reduced the number of kindergarten places. Tree-felling along the rivers leads to increased temperature and silt content. The regulation of water systems and climate change are altering water levels and temperatures alike. In short, the caretakers of the water systems face multiple challenges and the result is that hardly any small mussels are to be found in a third of the Norwegian rivers where freshwater pearl mussels live, while recruitment is poor in many of the rest.

Fortunately, there is hope. In Hordaland, western Norway, 'foster homes' have been set up for small mussels, where they live in a fully furnished residence with clean water at an appropriate temperature until they are big enough to look after themselves. Then they are returned to the river they came from. So far, the project has been a success. So maybe we can still foresee a future for the freshwater pearl mussels, the caretakers who clean up our waterways.

Poisoners and Purifying Moss

Arsenic has been the poisoner's favourite weapon throughout human history. Less well known is the fact that arsenic can be found in drinking water – and that a tiny tuft of moss can cleanse this harmful water, making it pure enough to drink in just an hour.

Potable water poisoned with arsenic is a health problem in several places around the world, especially in parts of Southeast Asia. Despite the best intentions, things went badly wrong here when UNICEF invested large sums in digging wells in villages in the 1960s and 1970s to secure clean water for the inhabitants. What nobody realised –

because arsenic is odourless, tasteless and colourless – was that arsenic was leaking from the rock and making the water poisonous. Only after millions of people showed visible signs of arsenic poisoning, and also displayed above-average incidence of cancer and other diseases, was the connection made. More than 100 million people, maybe as many as 200 million, are currently exposed to arsenic in water at levels that exceed the threshold values set by the World Health Organization (WHO).

In Skellefteå, northern Sweden, arsenic-rich minerals have been exposed by mining operations and the arsenic is leaking out into both surface and drinking water. Skellefteå is among the richest regions for mineral resources, and they are easily accessible for mining, so the Swedes have mined gold, copper, silver and zinc there for close to 100 years. It was here, recently, that a botanist conducting a field study noticed a delicate green moss that appeared to thrive in the arsenic-rich water. It was a species of hook-moss called *Warnstorfia fluitans*, whose long threads somewhat resemble green intestines as they bob on the surface of the polluted wetlands. Specimens were taken to the lab, where it was found that this floating hook-moss from Lapland was a total badass when it came to sucking up arsenic. Arsenic occurs in various forms, which are absorbed and accumulated in the moss, thereby reducing the arsenic content in the water and making it safe to drink. At low arsenic concentrations, it took just one hour for 80 per cent of the arsenic to be removed and for the water to be clean enough to drink. When the arsenic concentration was higher, it took longer, but the effect of the moss was equally impressive.

If only my late great-great-great-great-grandfather's brother, a wholesale merchant called Niels Anker Stang of Halden, had known about this. Both he and his wife were killed with arsenic, served to them – mixed in coffee and barley soup – by a serving girl called Sophie Johannesdotter. Her motives for the murders are unclear. It is said that Sophie fell out with the lady of the house and took her life first. Two years later, my distant relative discovered that the servant girl had stolen property from the house. Then it was his turn to be served arsenic.

The murders were only discovered when the Stang family home burned down a couple of months later, and it became clear that Sophie was the person who had set it on fire. She didn't just confess to the act of arson, but also to murder. The discovery of arsenic in the graves left little doubt about the cause of death and Sophie Johannesdotter was sentenced to death and beheaded in Halden at 9.30 a.m. on Saturday, 18 February 1876. She was the very last woman to be executed in Norway. The local newspaper report in Gothic script makes for eerie reading. Thousands of onlookers turned up, even though the time of the execution was kept secret. The axe fell during the fifth petition of the Our Father, 'Forgive us our sins', whereupon 'the executioner laid the head by the side of the body and the priest finished intoning the Our Father.' Johannesdotter was buried in an unmarked grave in a corner of Os cemetery, in Halden.

To get back to drinking water and the moss that cleanses it of arsenic: we know of only a few plants that can tolerate and absorb large quantities of arsenic, and this particular species of hook-moss is the only one that thrives in water. Since arsenic pollution in water is a public health problem, this kind of research is interesting and clearly relevant. A lot of research remains to be done before we will have the potential to set up purifying wetlands of *Warnstorfia fluitans*, and this particular moss can hardly be used in Asia – the climates of Lapland and Bangladesh aren't exactly identical.

Even so, the role of this hook-moss is a good example of a technique known as phytoremediation, where plants are used to purify polluted water, soil or air by tapping into their capacity to absorb and store, or break down, various harmful substances. This onsite plant purification may be an environmentally friendly – and cheap – alternative to mechanical or chemical methods. In both Europe and the

US, phytoremediation is now being tested out in field studies, to clean up after oil drilling, mining and many other types of polluting activities.

CHAPTER 2

A Gargantuan Grocery Store

O wonderful! I am food! I am food!
I am the eater of food!

The Upanishads,

Ca. 600 BCE

In the northwestern US state of Oregon, a minuscule organism has won great glory. In 2013 brewer's yeast – a tiny fungus with a waistline of around 5 micrometres (a tenth of the breadth of a strand of hair) – was designated, to great fanfare and with a resounding majority, the Official State Microbe. To the best of my knowledge, no other microorganisms share this honour. One important motivation was Oregon's brewing traditions, for which this species is crucial. But the microorganism also contributes to wine, sake, wholemeal bread, pizza bases and the enticing aroma of fresh currant buns.

Other species that help give us food and drink are more visible and better known: grains and rice, fruit and vegetables, meat and fish, as well as other seafood. Whether you opt to pick your berries in the forest or buy them from the heaving shelves of the supermarket fruit-and-veg aisle, nature is where they come from. Yet it is far from inevitable that these goods will continue to exist in the quantity and quality we wish for. Especially considering the way we humans have been trampling around in nature's supermarket without paying much attention to the massive footprints our boots are leaving.

Something Brewing – Wasps and Wine

Now and then, you may drink something other than water. Wine, for example. But before you'll be able to raise your glass in a toast, a whole slew of helpful species come into play. Vineyards with grapes are needed. Brewer's yeast is also crucial, even though you'll never see it portrayed on the label of a wine bottle in between the chateau and the vintage. But most curiously of all, it has recently been shown that social wasps also play a role in the production of good wine.

Varieties of brewer's yeast are needed to make fermented drinks, where they transform sugar and starch into carbon dioxide and alcohol. These sorts of drinks, in local variants, have been produced in societies throughout the world for at least the past 10,000 years. There are many reasons for this. Alcohol has disinfectant, analgesic and preservative properties – not to mention the effect it has on our personalities, which hardly requires description. The earliest documentation of winemaking comes from China, roughly 9,000 years ago. Nowadays, almost 30 billion litres of wine are produced worldwide per year, equivalent to the combined volume of 12,000 Olympic swimming pools.

Brewer's yeast is found in ripe grapes. But where does it come from? It's not present in unripe grapes, and in the other places where it has been found in nature – oak bark, for example – it is only ever present in the same warm summer months. Where is it the rest of the year – and how does it find its way into ripe grapes?

The answer is the social wasp 'in a striped bathing-suit, blazing and resolute' – as Norwegian poet Inger Hagerup put it – and on its way, belly full of yeast, to make fine wine. New research has shown that brewer's yeast lives in the wasp's stomach all year round; it is found in both the European hornet and the most common genus of paper wasp, which are close relatives of those yellow-and-black pests we hate so much. These social wasps provide both accommodation and taxi services: the wasp's innards offer the brewer's yeast a safe, cosy haven when weather conditions make the outdoors less appealing. The yeast can pass from mother to child because the adult wasps regurgitate what they've eaten themselves to feed their larvae. This allows the brewer's yeast to pass through the generations. What's more, some wasps – newly mated queens, to be precise – hibernate. After that, the queen and her daughters fly the brewer's yeast onward to next summer's grapes when they help themselves to sweet grape juice.

Different wasps convey different variants of brewer's yeast, which contributes to the unique qualities of the wine. Because the wasp belly isn't just a space where the brewer's yeast can cling on tight in some dark corner, waiting to be transported to a grape near your local wine producer. Far from it. It's all happening in there – for the yeast, it must be almost like going to a nightclub: down there in the muggy gloom of the wasp's belly, different varieties of brewer's yeast mingle, in their brewer's yeast way. As a result, new genetic variants come about, and each of these different variants helps flavour your wine in its own particular way.

So next time you're treating yourself to a drop of your favourite wine, raise a glass to the inner lives of insects.

If You Are What You Eat, You're Walking Grass

Grass

I grow in places
others can't,

where wind is high
and water scant.

I drink the rain,
I eat the sun;

before the prairie winds
I run.

I seed, I sprout,
I grow, I creep,

and in the ice
and snow, I sleep.

On steppe or veld
or pampas dry,

beneath the grand,
enormous sky,

I make my humble,
bladed bed.

And where there's level ground,
I spread.

JOYCE SIDMAN

from *Ubiquitous: Celebrating Nature's
Survivors*, 2010

Did you think only cows ate grass? Think again. Because almost half the calories we humans consume come from grass plants.

Most of what we eat belongs to the plant kingdom. This is hardly surprising: plants are literally the foundation of all other life. Unlike them, we humans can't produce our own food based on lifeless substances such as carbon dioxide and water. This is the magic of photosynthesis: the plants 'eat' sunlight, carbon dioxide and water, conjuring them into organic molecules – living plant biomass. And their achievements are hardly trivial. Every year, the primary producers of the planet (plants, algae and some bacteria, to be precise) extract a good 100 billion tonnes of carbon from the atmosphere, combining it with other elements to create everything from stalks of grain to massively heavy giant sequoia trees.

All the rest of us organisms on the planet are, either directly or indirectly, 100 per cent reliant on the work plants do to obtain the energy and building blocks we need to construct our own bodies (with the exception of some strange deep-sea communities that live off chemical energy). Of course there is another thing plants produce too, as a kind of waste product of their photosynthesis: that not-insignificant gas, oxygen.

There are at least 50,000 edible plant species on the planet but humans have only cultivated an estimated 7,000 of these as food sources at some point over human history. Nowadays, that number is far lower – between 100 and 200 – and a few plant species are becoming increasingly dominant. Rice, corn and wheat alone now account for almost 60 per cent of the calories we obtain from the plant kingdom. One unhappy consequence of this is that the wild relatives of our food plants, which have genetic material we can use to develop more robust food plants, are in retreat – a fifth of them are threatened with extinction. Unfortunately, although agricultural crop yields have increased since 1970, thanks to artificial fertiliser and chemical pesticides, nature's capacity to shore up food production has been diminished. This is because services like pollination and natural pest and weed control have been reduced over the same period, according to IPBES. (The Intergovernmental Science–Policy Platform on Biodiversity and Ecosystem Services is sometimes called 'the sister of the IPCC', the Intergovernmental Panel on Climate Change, and is an independent, intergovernmental body that brings together global scientific knowledge in the field of the environment.)

Whereas cows chew on stems and leaves, we humans eat the offspring of the grass: the seeds. Rice, corn and various types of grain are simply grass seeds. These contain many carbohydrates in the form of starch, which are designed to be a packed lunch for the tiny new shoots, and we are good at digesting starch. But we have a tougher time of it with cellulose, which accounts for much of the rest of the plant and is the substance we use to make paper.

Some days, I may be so busy or having so much fun at work that I totally forget to eat. On one such day in Oslo, full of meetings that involved intense discussions about measuring the state of Norwegian nature, I didn't get around to starting on my homemade packed lunch, a tasty double-decker cheese roll, until I was heading for the tram at a half-run. My thoughts were still a long way off, occupied with blueberry cover and reference conditions and dead wood in forests. Only as I was about to take the last bite did I realise that I'd gobbled down not just the food, but also the paper separating the two layers of my lunch. A happy union of starch and cellulose, then.

The problem for us humans, if we happen to try eating cellulose, is that we lack the biological agents – enzymes – capable of pulling apart the strong bonds and making the nutrients accessible. There is, in fact, no vertebrate capable of digesting cellulose. Cows can't either, or at least not without help. But cows' stomachs accommodate roughly 3 kilograms of bacteria, fungi and single-cell organisms, some of which are able to release the nutrients from grass and hay (as to whether or not they can also deal with sandwich paper, no comment). We humans also have rich gut flora, but we don't have any capable of breaking down cellulose on our team.

If you now find yourself asking why we can also eat parts of plants other than grass seeds – like spinach and salad, root vegetables such as potatoes, and fruits and berries – it's because they have a relatively low cellulose content and more easily accessible nutrients, like starch. I'll come back to fruits and berries, plant-based food that provides us with important vitamins and minerals, in Chapter 3 on pollination. For the time being, let the following facts sink in, since we've spoken so much about grass: did you know that the most water-intensive 'crop' in the US is, pointless as this might seem, lawn grass? The country's lawns and golf courses are estimated to account for 1.9 per cent of the US, and this grass requires more watering than American farmers expend on corn, rice, fruits and nuts put together.

An Avalanche of Extinctions –
The Megafauna That Vanished

So, an average person on planet Earth today acquires most of their energy from the plant kingdom: more than 80 per cent of humanity's calorie intake consists of various kinds of grains and produce. The remaining calories come from the animal kingdom – roughly a tenth from meat, including animal fat and organs, and the rest from eggs and milk, and seafood. The meat we eat also comes from nature – although this is no longer strictly true now that much of the world's meat production looks more like industry than nature.

But let's zip back in time. If you could take a look in history's rear-view mirror, you'd see that our planet was home to extremely large animals for millions of years ... until we meat-hungry humans made our entrance.

Think of the word 'camel'. What do you picture – caravans winding through the Sahara, perhaps? Cud-chewing creatures in the barren landscape of the Gobi Desert? But did you know that the camel originated in North America? Up until the end of the last ice age, around 12,000 years ago, *Camelops hesternus* (also known as 'Yesterday's camel') ambled its way through the places where Los Angeles and San Francisco now stand. Camels spread to Asia via the Bering Strait before and at the same time as Yesterday's camel died out in North America.

Or take elephants. If we go back a few hundred thousand years, animals with trunks – proboscideans – lived all over the world, with the exception of Australia and the Antarctic. Straight-tusked elephants, mastodons, mammoths. On the holiday islands of the Mediterranean there were small dwarf elephants around a metre in height, roughly the size of a Shetland pony. As recently as 14,000 years ago, the mammal fauna on the American continent was more varied than that of modern-day Africa. And then the human race expanded, rose up on its hind legs and took a firm grasp on its spear in a world where it was surrounded by great beasts.

Over a relatively short period, just dozens of millennia ago, more than half of the megafauna species vanished. Gone were the sabre-toothed tiger, the giant sloth and the Irish deer, the American cave lion and Europe's woolly rhinoceros. What happened? This is a source of much heated debate, but there is little doubt that hunting by humans played a role, probably in tandem with climate change.

On continent after continent it is striking to see how the disappearance of the megafauna coincides with the arrival of humankind. The effect becomes clearly visible if we compare the average size of existing mammal species at different points in time. Following the arrival of modern humans in Eurasia, the average size of the mammals they shared the land with halved. In Australia, the average size of mammals fell to a tenth after humans reached the shores of the island continent sometime between 40,000 and 60,000 years ago. Most dramatic of all is the pattern in America: 10 per cent of all mammals that existed there when Stone Age humans came wandering across the dried-out Bering Strait with their spears vanished. Here too the worst affected were the largest species: all animals weighing more than 600 kilos died out and the average weight of North American mammals fell from 98 to 8kg.

To sum up this almost unfathomable phenomenon: of the at least 50 herbivorous mammals weighing over a tonne that existed in the world 100,000 years ago, only nine remain, and of the 15 large (over 100kg) predators that existed then, just six have survived.

These changes had profound effects on the entire eco-system, which we have barely begun to grasp – not just an avalanche of extinctions of other species whose traces are less easy to see, but a reshuffle of the entire food chain and ecological processes. To cite one of many scientific articles published in the past decade: 'Only recently have we begun to appreciate just how drastically the Earth has been modified by human activities and for how long'.

Most important of all, perhaps, was the effect these giant beasts had on even the physical structures – that is, the appearance of the ecosystem – whether dense forest or a half-open savannah landscape. From studies in Africa, where we can still investigate the ecological effect of a megafauna that remains somewhat intact, we know that big animals can reduce forest cover by 15 to 95 per cent by damaging or toppling large trees and trampling smaller ones. Studies from Kruger National Park in South Africa show that an African savannah elephant can topple up to 1,500 fully grown trees every year. It is likely that other parts of the world – such as South America – would also have had more open savannah had the megafauna survived.

The extinction of megafauna also had ripple effects on the species that were eaten by them, lived on them as parasites, ate their dung or carrion, or used them to disseminate their seeds. To take one concrete example that is easy to relate to: when the giant sloth in Latin America died out roughly 10,000 years ago (it was big and slow and probably easy to hunt and, being the size of a VW Beetle, would have provided meat for many people), the avocado lost its seed-spreader. Avocado trees grow in hot, humid forests where they produce their familiar seed cushioned in soft, pale-green flesh and enclosed in the skin. The giant sloth was enough of a bigmouth to swallow the whole of this glorious green fruit in one gulp. The flesh disappeared

The giant sloth, which became extinct roughly 10,000 years ago, used to eat avocados and helped spread their seeds.

into its digestive system and the seed emerged from its backside, ready-planted in a heap of nutrient-rich manure.

Despite the 10,000 years it has had to ponder this, the avocado still doesn't appear to have grasped that its seed-spreading pal is gone. Wild avocado trees still produce their large-stoned fruits and in nature, large numbers often end up lying around beneath the mother tree, competing with each other for light and water and nutrients. Unless a human turns up, that is, because we have taken over the giant sloth's role as avocado eaters. The Aztecs had already been eating avocados for centuries by the time the Spaniards arrived in the early sixteenth century. Indeed, its name comes from their language: they called it *ahuacatl*, which means testicle, apparently because they grow in pairs in a way reminiscent of that particular body part.

Next time you eat guacamole, shed a tear for all the other tasty fruits and vegetables that may have died out along with the megafauna without us even getting to know about them. If only they'd survived, perhaps avocado toast would have greater competition on the menus of hipster cafés.

Meat-hungry – Past and Present

We don't know how far back in time our forebears' history as hunters stretches, but we are talking millions of years. For almost all of this period, we have bagged our prey using simple methods: driving animals off cliffs or into trapping pits, and using sharp or heavy objects like rocks or spears to kill them. While these methods were technically simple they were – as described above – surprisingly effective all told. But they required a detailed knowledge of the hunted animal's behaviour and habits, and of nature in general; knowledge that is well on its way to being lost today.

One interesting example of the intimate understanding of nature that it takes to successfully bag one's prey is still – just – to be found in southern Africa. In the Kalahari Desert live the San people (also known as Khoisan or 'Bushmen'), a common term for tribes that use languages with clicking consonants. Some tribes among the San people still keep an ancient hunting tradition alive. They hunt with bows and arrows whose tips are poisoned with insect venom.

Shooting with bows originated pretty late in human history, apparently coinciding with notable changes in social organisation some 70,000 years ago. It is difficult to document but scientists believe the use of poison-tipped arrows or spears was already important in prehistoric times.

The word *toxin*, i.e. poison, comes from the Greek word for arrow and/or bow: *toxon*. We are familiar with the use of poisoned arrows both from indigenous peoples (the use of curare as an arrow poison in South America, for example) and from myths and legends. Odysseus poisoned his arrows with hellebore. And in Norse mythology, kind Balder was killed by an arrow made of poisonous mistletoe – although the poisonous nature of the plant is not the point in this story, it is rather the fact that mistletoe was left out when Freya made all things living and dead promise never to hurt her fair son.

The San people do not use hellebore or mistletoe. They find a myrrh, a low-growing tree species whose resin has a sweet aroma – yes, the same myrrh you've heard about in the Nativity story – and dig a deep hole down to the roots. Down there lie the larvae of a leaf beetle, each in its own tough cocoon – a kind of homemade sleeping bag of earth and faeces. The cocoons are collected and opened. When the larvae are squeezed, like a tube of lip balm, poisonous gloop oozes out. This venomous mixture is applied to the arrow tip with a little stick – and then it's ready for the hunt.

We're not talking about small game hunting here, either: everything from giraffes to elephants is hunted with poison arrows. The poison doesn't kill the prey immediately but does affect its oxygen transport. It works primarily by dissolving the red blood cells. Since these are what carry the oxygen around the body, the animal will die slowly by a kind of internal suffocation. Meantime, the San hunters follow their prey. The hunt may last hours, days. Physical endurance and toughness are crucial, as well as good tracking skills.

As with other types of arrow poison, the venomous substances only work if they enter the body via the blood. Consequently, it isn't a problem to eat an animal killed with this kind of poison – although it is dangerous, for the hunters too, if the venom enters a wound.

Different tribes have different variants of this hunting tradition. Some use other species of larvae, others mix their poison using certain poisonous plants. Nowadays, these traditional hunting methods are prohibited in most places where the San people live. As a result, the opportunities to observe and learn about and from this culture – including knowledge about which animals and plants are edible, poisonous, etc. – are also in the process of being lost.

Even using these sorts of 'primitive' hunting methods, early humans still managed to leave a deep impression, as we have seen. Yet the extinction of the megafauna in the Stone Age was just the beginning. The effects of our quest for food have accelerated in line with our rising population and the inventions of modern civilisation – a topic for a whole separate book. Here, we only have space to look at broad patterns in the way our food production affects nature today.

Global per capita meat consumption averages 44kg a year, equivalent to the meat weight of four lambs. That's almost twice the level in the 1960s, the decade of my birth, and our meat consumption has major consequences. Half of the entire land area on Earth that isn't ice or desert is used for agriculture, but only a fifth of it is used to cultivate the end product, i.e. human food. The rest is given over to our domestic animals, either for grazing or the production of fodder.

Nowadays, almost all of our animal protein comes from domestic animals: we've filled the gap left by Dumbo's wild trunked forebears with Daisy the cow. The combined weight of our domestic animals today is more than 10 times as high as the estimated weight of the megafauna living in the wild before the Stone Age. Our poultry alone weigh almost three times more than the weight of all the world's wild birds put together. In addition to the ecological challenges, this poses a whole series of ethical and animal welfare problems. Cutting meat consumption in the most meat-hungry parts of the world is a simple, environmentally friendly contribution to more sustainable food production.

The Sea – The Last Healthy
Part of a Sick World?

It is easy to forget the sea, despite its vast size: it covers more than 70 per cent of our planet and has an average depth of around 3 kilometres. Ninety-five per cent of the ocean bed has never been seen by human eyes and we have more accurate maps of the surface of Mars, which is separated from Earth by an ice-cold vacuum of 262 million kilometres. Even so, nobody could deny that the ocean provides us with vital natural goods and services: not just fish and seafood but salt and seaweed to wrap our maki rolls in. The ocean also contributes to support services, such as circulation of nutrients, climate regulation and the water cycle, not to mention oxygen production – you can thank the ocean's green plankton for at least every other breath you take.

But how much do you know about the fish on your plate? About where it and the world's other edible fish come from? I was certainly surprised when I began to look into global figures and learned that more than half of all fish for human consumption now come from fish farms and more than half of those again are raised in freshwater. This shift is a response to both changes in ocean fish stocks and our increasing demand for fish.

On a global basis, we eat around 20kg of fish per head per year. When I was born, the figure was around 10kg, so consumption has ticked slowly and steadily upwards. The diets of populations in developing nations contain a larger proportion of fish than in industrialised countries and, as with meat, the consumption span is wide – from more than 50kg per head in the tiny island states of the Pacific Ocean to 2kg in Central Asia.

Like hunting, fishing and the hunting of marine mammals have long traditions and considerable ecological consequences. The UN Food and Agriculture Organization (FAO) reports that a third of the world's fish stocks are overfished. We have not just altered biodiversity but the ocean itself as an ecosystem. As with the land animals, many of the largest species have suffered the brunt, such as predatory fish (sharks, stingrays and swordfish) and whales, and this has affected the entire food chain – in a ripple effect.

≥◦≤

In the log cabin of my childhood, my sister pinned a piece of paper up over the bed. It bore a quotation from the Norwegian author Alexander L. Kielland (1849–1906): 'It is untrue that the Sea is faithless, for it has never promised anything: no demands, no obligations, free, pure and true the great heart beats – the last healthy thing in a sick World.'

I'm not so sure about that, sadly. The ocean simply isn't healthy any longer, what with acidification, oxygen-free dead zones, microplastics – and overfished stocks. The first three challenges are new. But it is far from certain that the oceans were healthy in the 1880s either, when Kielland wrote *Garman & Worse*, because (predatory) fishing is hardly a new invention. The examples we know of, based on historical sources, are dramatic. Like a study of great fish species in the North Atlantic, which estimates that the combined weight of the great fish (more than 16kg) is now less than 3 per cent of what it would have been without fishing. Or the comparison of cod stocks off Newfoundland in 1505 and 1990, which estimates that 99 per cent of cod has vanished.

Of course, it is difficult to document changes in fish stocks when we lack any solid, systematic database that extends far back in time. These are uncertain estimates but they give the distinct impression that things are not as they should be down in the briny blue.

To finish up where we started off, more than 4,800 people have reached the highest point on Earth – Mount Everest. Fewer than 600 people have been to outer space, and 12 have even walked on the moon. But only four people have visited the deepest place on the planet, the Mariana Trench. We must accept that it is impossible to have a full overview of millions of cubic kilometres of saltwater and everything that lives in it, but that does not mean that the oceans can cope with absolutely anything.

We must take greater pains to make the ocean 'pure and healthy'. Fish and seafood farming (both at sea and in freshwater) helps provide a growing human population with protein, but far from all facilities comply with the stipulated environmental standards. In order to harvest seafood sustainably in the future, further work must be done on many fronts: carrying out proper quota calculations, halting illegal fishing, reducing bycatch, setting up marine conservation areas. We must try to save the coral reefs, which are under threat from warmer seas, acidification and pollution – they support more species per area than any other ocean ecosystem. Although the reefs account for less than 1 per cent of the world's oceans, at least a quarter of all marine species are estimated to spend part of their lives on a coral reef. All this and more we must do, even if we don't have full oversight of the resources in the deep blue sea, to ensure that we do not end up with the scenario described by distinguished fishery scientist John Gulland: 'Fisheries management is an endless argument over how many fish are in the sea until all doubt is removed, but so are all the fish.'

Shifting Baseline Syndrome: Why We Don't Notice Deterioration

does the butterfly remember

what the caterpillar knows?

ANJA KONIG

Excerpt from 'Metamorphoses', from the

collection *Animal Experiments*, 2020

More than 30 years ago, I was driving through the southern states of the US, all the way out to Key West, the country's southernmost point. The Overseas Highway – the road from mainland Florida through the Keys – stretches across the ocean, each straight segment linking one island to the next, as if in a dot-to-dot picture where the pattern doesn't appear until you've connected all the dots with pencil lines.

Other than watching the famous sunset (and getting a rude awakening from the police, who scolded us for staying overnight in the car), one of my clearest memories of the trip is Ernest Hemingway's house, with his typewriter and all the six-toed cats – descendants of Hemingway's ship's cat, which had this elegant little genetic defect. Hemingway was also fond of fish, or, more accurately, of catching them. A photo from 1935 shows him posing, tanned and smiling, with his family in front of four dangling marlins almost twice as long as he was tall.

On Key West, it has long been traditional for people to take a trophy picture of the day's catch. If you take out these historical photographs and place them in chronological order, you will see a clear pattern: in the newest photos the angler's smile is just as broad as in the older ones, because he thinks he has caught a gigantic fish, but the trophy fish has shrunk. Between 1956 and 2007, the day's biggest catch shrank from 92 to 42cm or – in estimated weight based on length and species – from 20kg to a measly 2.3kg. Yet if you asked the person who hauled ashore the record fish in 2007, he would probably answer: Oh yes, his was the very biggest fish anybody had ever seen. Because we suffer from a kind of collective amnesia.

Can you describe something you don't remember? A state of nature that you haven't seen? Hardly. And that is the very core of a psychological phenomenon that relates to nature and is known as 'shifting baseline syndrome': a shifting, altered baseline for the state of the natural world.

The phenomenon describes how, over time, we lose knowledge about nature's 'health status' because we do not comprehend the changes that are actually happening. Our short lives and limited memories give us incorrect impressions of how profoundly the world has been altered by our activity because our baseline changes with each generation. Sometimes even in the course of a single generation. It's like the way you're sure you must have misremembered

The marlin is an attractive species in offshore game fishing and is often displayed as a trophy fish.

when you think back to how much cod you used to be able to fish in the Skagerrak or indeed elsewhere as a young kid. Gradually, you reduce your expectations about the natural world around you.

A Canadian marine biologist was the person who coined the term 'shifting baseline'. Daniel Pauly was studying the way we first overfished the big predatory fish and then, when there were too few for the fishing to be profitable, shifted our focus steadily down the food chain. In addition, he made a crucial observation: he saw how both fishermen and marine biologists interpreted species and numbers based on memories from early in their careers – as if this were an unaffected reference point, which they then used as a benchmark for gauging new changes. As a result, they accepted ever-more impoverished ecosystems as the normal state of nature. This is shifting baseline syndrome: a kind of shared amnesia about the state of nature from one generation to the next.

Let's take another example, a fish that moves between sea and freshwater. The amount of salmon in the Columbia River in the US is twice as high today as it was in 1930. This sounds great – if 1930 is your benchmark. However, in 1930, salmon numbers in this river were just a tenth of the level seen in the 1800s. *That* benchmark gives a totally different picture of the long-term changes that have taken place – and thereby, also, a different basis on which to grasp the impact of the change.

Shifting baseline syndrome is not some sentimental romanticism that claims 'everything was better in the good old days'. Nor is it a naive belief that we should return to the state of nature, living like Stone Age people in a vast, untamed natural world. Consciousness of shifting baselines is, rather, about having a proper starting point for our calculations when we are taking stock; using the right exchange rate when we are attempting to assess where the planet's limits lie.

Humanity's staggering capacity to adapt is a strength but also a weakness. Our collective amnesia prevents us from grasping just how much we have altered nature, because we continuously get used to the new normal – whether that's fewer squashed insects on our windscreens, the absence of old and dead trees in the forest or ever more frequent extreme weather. This also makes it more difficult to get most people and politicians to understand the gravity of the situation and become involved. At a time when the Earth's ecosystems are dwindling at an ever-increasing rate, our shifting baselines are a major challenge.

CHAPTER 3

The World's Biggest Buzz

One of my earliest memories involves coffee. As a pigtailed and snowsuit-clad four-year-old, I was once foolish enough to lick the metal railings around Bamsebo playground. It was an ice-cold winter's day in Northern Norway and my tongue froze to the spot. I was rescued by a day-care worker who ran up and poured a mug of coffee over my frozen-stuck tongue.

Fifty years on from this liberating first encounter with *Coffea arabica*, I remain a firm fan of coffee. And I'm not alone: a billion cups are drunk around the world every single day and the beverage has sparked anger and accolades alike throughout history. The English king Charles II tried to ban the serving of coffee in 1675 because he considered coffee houses to be hotbeds of rebellion-minded intellectuals. Half a century later, the composer Johann Sebastian Bach wrote his popular secular work, the 'Coffee

Cantata' in which a young girl pleads with her father for permission to try the new, fashionable drink ('*Kaffee, Kaffee muss ich haben ...*').

Imagine a daily life without morning coffee. Or Saturday nights without chocolate, or Christmas celebrations without marzipan. Not to mention our Norwegian tradition of Taco Friday without taco shells or nachos (because they contain sunflower oil), and with sweetcorn as the only vegetable. This could be our new normal if we fail to look after the world's insects and enable them to pollinate our food plants, because fruits, berries, many vegetables and nuts are largely pollinated by insects and cannot be cultivated without the help of wild insects – at least not on the same scale or as cheaply as today.

Insect-pollinated crops don't just bring flavour and colour to our plate, they are also a source of vitamins and micronutrients. Consequently, the ongoing decline in pollinators may also be detrimental to our health – especially in parts of Southeast Asia, where half of all plant-based sources of vitamin A rely on pollination by animals.

In fact, pollination is also a matter of justice and solidarity. People in Laos, where a third of the population live below the poverty line, won't necessarily be able to compensate for the lack of nutrient-rich plants by taking vitamin pills. And many of the food plants that rely on pollination are important sources of income for small-scale farmers and family-owned farms in developing nations. Insects provide the basis for the jobs and incomes of millions of people – that's why the sound of insects is the world's biggest buzz.

The Blossoms and the Bees

Understanding pollination requires a spot of sex education; a somewhat expanded version of the birds and the bees – or, as we say in Norwegian, the *blossoms* and the bees. Let's start with the blossoms. Since plants are tethered to the ground by their roots they've had to find a different way of reproducing from us animals. A tomato plant or apple tree can't exactly run around looking for a suitable partner so in order to secure the continuation of their line, plants switch between two different generations, one of which is, in fact, able to move around – with a bit of help.

One generation is clearly visible: a green growth with stem, leaves and petals. This is what you think of as a plant, which is hardly surprising since it both grows largest and lives longest. This growth then produces the next generation: small, short-lived individuals, which are male or female. The female individual is neatly packaged up in the plant's flower; the male individual is also produced in the flower and is what we know as pollen. It isn't much of an individual. This little speck is more like a kind of plant penis, full to the brim with genetic material and protected by an extremely resistant shell (*more on this in Chapter 8, see page 194*). A functioning sex life requires this tiny mote to find its way to a female individual, preferably on another plant, so that they can exchange and combine their genetic material. And this is where they need assistance.

A plant that pins its hopes on the wind will need to produce massive amounts of pollen if it's to have any chance of being carried away and ending up in another flower of the same species. We're talking about hundreds of millions of grains of pollen from a single plant. For those with pollen allergies, this is an example of a benefit of nature that is not entirely positive. Conifers, many plants with catkins, grasses and grass-like plants rely on this mass-production strategy. They have small, insignificant flowers that are generally green.

Other plants produce huge, brightly coloured flowers and rely on assistance from the animal kingdom to transport the pollen grain to its date with the female individual. In this case, insects are a particularly important freight firm. Insects and flowering plants have evolved in tandem since the Cretaceous period more than a hundred million years ago. It started out as a handy coincidence: a hungry beetle in search of breakfast happened to find a magnolia flower with delicious, nutritious pollen hidden amid its thick petals. While it was gobbling down both pollen and flower, some of the pollen got stuck to the beetle's body. It flew onwards, found a new magnolia blossom and voilà – the pollen from the first flower pollinated the second flower and reproduction was secured.

Evolution continued on its way and soon the bees arrived: specially developed flying birth assistants for flowering plants. There are many kinds of bee, around 20,000 species worldwide, more than 200 of them in Norway. That includes close to 30 social bumblebee species that live in colonies, seven cuckoo bumblebees that take over the nests of the social species and lay their own eggs there, more than 170 solitary wild bees (in other words, bees that don't live in colonies but on their own, like most insects) and our own six-legged livestock, the honeybees. Honeybees give us honey and contribute to pollination, but all the other insects combined are responsible for the lion's share of pollination.

Bees have a couple of things in common that make them eminently well-suited to their role as pollinators. First off, they are hairy and, as if that wasn't enough, their hairs are branched. When magnified, they almost look like tiny feathers, as each hair may have many side branches. This makes it especially easy for tiny pollen grains to attach themselves all over the bee. Secondly, bees are vegetarian. They live solely on pollen and nectar. The bee young – the larvae –

are bottle-fed on this diet. In the case of wild bees, the female collects the pollen and nectar, kneading it into clumps that are placed with the eggs. Since bees need to find a lot of pollen and nectar to feed their larvae and themselves, they make a great many flower visits and do plenty of good pollination along the way.

Given the enormous contribution of wild bees and other pollinating insects, they cannot be replaced by domesticated honeybees. This has been demonstrated across many different food plants and crops worldwide. For example, an American study of apple cultivation shows that almost 1 per cent more flowers developed into apples for every additional wild bee species present, whereas the presence of honeybees did not yield more apples. One of the reasons for this may be that, whereas wild bees diligently flew to all the apple trees, the honeybees preferred to cruise straight to the ones with the most blossom.

The global value of directly pollination-dependent food production is somewhere around two-thirds the size of the UK's 2016–17 government spending. The volume of agricultural plants that require pollination has tripled over the past 50 years but crop yield has failed to increase at the same pace over the same period. One explanation for this may be the fact that pollinating insects appear to be going in the opposite direction: in many places across the world, recent studies have shown a constant decrease in the number of individuals and a reduction in the diversity of our tiny flying helpers.

Blue Honey Makes Beekeepers See Red

Several of the social bees produce some honey but only the honeybee species does so in such large quantities that we can harvest what they have diligently gathered. And it is quite a job: producing a kilogram of honey entails visits to several million flowers. It may be tempting to opt for quick and easy solutions, and a lot can go wrong in encounters with the modern world. A few years back, for example, beekeepers in northeastern France got the shock of their lives when they peered into their hives and saw that the honey in the wax honeycombs wasn't the usual warm golden colour but blue or green instead.

The beekeepers saw red, because although the honey tasted fine it was impossible to sell. And the dozens of affected beekeepers felt they already had enough problems on their plate, what with bee diseases and low honey production. They embarked on a painstaking round of detective work. First, they noticed that the bees were returning to the hives with an unidentifiable brightly coloured substance in their pollen baskets. And then they tracked down the source. It turned out to be a biogas facility some kilometres away, which was storing waste material from a factory that produced brightly coloured chocolate-covered peanuts, M&Ms – outdoors and in the open. Perhaps the honeybees thought they had discovered some unusually large and nectar-rich flowers? Whatever the case, this was a stable, easily accessible source of sugar. There was no need for the bees to bother flying from one apple blossom to the next. Fortunately, the biogas facility moved all the raw material indoors as soon as the problem was identified. And with that, the French honey regained its golden hue.

This example illustrates why I am sceptical about putting out sugar water, banana skins and other stuff to help pollinating insects in the garden. This kind of thing may both distract them from the job they are supposed to be doing as pollinators and create an infection hotspot because too many individuals are visiting the same place. What's more, sugar water is a pale imitation of the real things that flowers produce – nectar and pollen – so it's far better to plant or sow the seeds of nectar-rich flowers in your garden. And please take note: never, ever give honey to insects. It may contain dormant bacteria that can infect honeybees and make them sick. Both American and European foulbrood can be transmitted in this way and both diseases are just as nasty as they sound.

Two Flower Flies with One Swat

Many species other than bees contribute to pollination: beetles, wasps, butterflies and moths, not to mention flies. In chilly climes especially – in high latitudes and way up in the mountains – flies are crucial. If you sit in the mountains near Finse (a station on the Oslo to Bergen railway line that is 1,222 metres above sea level) and check who's landing on the wild flowers over the summer, you'll find that more than eight out of 10 pollinators are house flies and their close relatives. They may lack the panda factor of the cute, fuzzy bumblebee, but these industrious insects are among our most important pollinators.

Flies also help out in temperate regions, especially flower flies. They're easy to spot because although disguised as wasps, in yellow and black stripes, they know a trick the wasps must envy: they can hover in the air, seemingly motionless, like miniature hummingbirds – hence their alternative name, hoverflies. They use this skill to suck nectar from flowers, but they also use their 'freeze' technique in a kind of hoverfly hover-off, where the male with the best moves is the coolest kid on the block and gets to mate with the female.

Every spring at least half a billion hoverflies come speeding across the Channel to the United Kingdom – because birds aren't the only ones to migrate, some insects do it too.

Butterflies and dragonflies are the best known of these but a study using radar showed that billions of hoverflies also migrate according to the season. And the UK's invasion by these hordes of flies is good news – very good news. Not only do adult hoverflies bear a cargo of exotic pollen from far-off places, they also provide an extensive inland transport service. On top of that, their young, hoverfly larvae, are gluttonous predators that polish off between three and 10 trillion aphids every summer, thereby helping protect our crops.

That is why hoverflies are also used as a natural form of pest control, as an alternative to spraying with insecticide. Two flies with one swat, in other words: pollination *and* pest control services in the same yellow-and-black-striped body. And fortunately, despite a growing number of dismal reports about declining insect populations, the numbers of these migrating hoverflies have remained stable for the past 10 years.

Brazil Nuts and Flying Perfume Flacons

Flower-pollinator relationships are often broad and not particularly specialised, enabling many different insect types to pollinate a given plant species. But in some cases, highly specific interactions have evolved that are so bizarre you'd barely believe them possible.

Let's take a trip to South America, where Brazil nut trees grow scattered throughout the rainforests. They can live for hundreds of years and tower as high as 40 metres in the air. From these heady heights, at a certain time of year, the tree's offspring come whizzing to the ground in the form of coconut-like capsules. These can easily weigh a couple of kilograms and, while fascinating, you certainly wouldn't want one landing on your head.

To quote Alexander von Humboldt, the German natural scientist who travelled in South America for several years in around 1800: 'These fruits, which are as large as the head of a child (...) make a very loud noise in falling from the tops of the trees. Nothing is more fitted to fill the mind with admiration of the force of organic action ...' Hidden inside the head-sized capsule are the Brazil nuts you find in bags of mixed nuts at Christmas time – the long oval ones that are impossible to crack without simultaneously crushing their contents.

I think von Humboldt and his travelling companion, French botanist Aimé Bonpland, would have been even more impressed by Brazil nuts had they known the antics required to pollinate the trees. The flowers of the Brazil nut tree are pollinated by creatures of almost unearthly beauty, whose shimmering bodies in shades of metallic blue, green and purple look like flying jewels. These insects are called orchid bees and they only live in South and Central

America. The females are in charge of pollen transport and it demands their full dedication, because Brazil nut flowers are sealed with a kind of lid and the female orchid bee is one of the few creatures able to squeeze past it and into the flower that conceals the nectar. This feat provides her with food and the Brazil nut tree with pollination, enabling nut production. But that is only half the story.

The orchid bee female is peculiar, you see. She will only mate with an orchid bee male that smells good. Since the male can't simply pop into a perfume shop to pick up a flacon of seductive scent, he has to make the perfume himself. So, while the orchid bee females are busy pollinating the Brazil nuts, he flies from orchid to orchid, gathering sweet-smelling oils, which he stores in a peculiar structure on his hind legs. A kind of triangular container formed by his leg plates – practically speaking, a perfume bottle.

This scent collection appears to be crucial for attracting orchid bee females. By producing his own unique *eau de parfum*, the male secures an opportunity to mate and produce new orchid bee young. At the same time, the male's flight from flower to flower ensures that pollen is transported among the orchids, enabling them to produce seeds.

So, the orchid bees' struggle to gather nectar and sweet scents benefits both the Brazil nut tree and the orchids – as well as us humans, by providing Brazil nuts for local populations and the export market. Once you've caught a glimpse of these complicated interactions among several species, you'll also understand why Brazil nuts can't be grown in plantations. Only in forests, where the living conditions of all the partners are assured, can bee, tree and orchid form their remarkable bond of friendship.

The Fig Tree and the Fig Wasp: Millions of Years of Loyalty and Treachery

Another example of extreme adaptation between pollinators and plants is that of the fig wasp and the fig tree. And it doesn't just involve friendship but an intimate partnership, complete with loyalty, self-sacrifice and betrayal. As they say in Hollywood romances: it's complicated. Just listen to this ...

Fig trees don't have normal flowers, the kind that open up to the world and offer themselves to any old flower-visiting insect. Nope. This story is strange from the start, because the fig tree's flowers are inside out. The trees produce small pale-green fruit-like pears that are hollow, and there inside the pear – or more accurately, the fig – are all the flowers.

It may seem pretty inept to hide the flowers away like that but the fig tree has a cunning plan. There is, in fact, a way in – a narrow passage – and this is where a suitable fig wasp makes its entrance: a female fig tree wasp that has

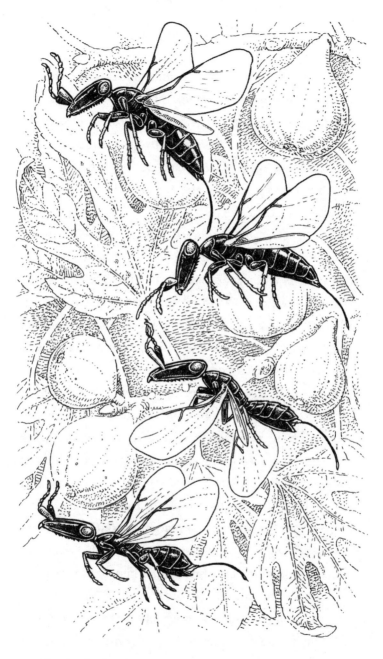

Pollinating fig wasps have a very special relationship with the strange
'inside-out' flowers of fig trees.

already mated squeezes her way in. The passage is so narrow that she loses her wings along the way, so she is trapped inside the flower cave for the rest of her short life. From the fig tree's point of view, that's perfectly fine. All it cares about is whether the fig wasp has brought along some pollen from another tree to fertilise the female flowers inside the inverted cluster of blooms.

The prospects of the trapped fig wasp look more dismal. For her, this is the lottery of life: has she crept into a fig that can serve as a nursery or has she been tricked into a relationship in which her fig tree partner will deny her children? Because – hang onto your hat, this is where things get complicated – it takes *two* varieties of the common fig to produce the species of fig we usually eat.

Some of the trees have functioning female flowers within the hollow, able to develop in such a way that the pear-like inverted cluster of flowers transforms into an edible fig. But the female fig wasp won't be able to lay any eggs in there, because of the physical structure of these functioning female flowers. If she has crawled into that type, she is a loser in life's lottery and will not get to continue her line. She has been betrayed by her fig tree partner, which has simply tricked her into bringing it her potent cargo of pollen.

Luckily for the female fig wasp and for this complicated relationship, common fig trees also come in a different variety, known as goat fig trees. Here, too, the entire blossom cluster is inverted, but the female flowers in the hollow are sterile and perfect for laying eggs in. There are also plenty of male flowers, which produce pollen. If the female wasp has drawn the winning ticket and crawled into one of these goat figs, the way is paved for a big brood. And when all the new fig wasps are ready for adult life, the fig tree is transformed from a nursery into a bordello, where the newly hatched wasps mate with one another.

Now there's just one problem left: how are the newly mated females to make their way into the world with their wings intact? This is where the male wasps come in. They may be blind and wingless, but they're champions at using their chewing mouthparts to expand the narrow tunnel, creating a roomy road to freedom for the females. On their way out, the female fig wasps pick up pollen from the male flowers and, as the male wasps die where they were born, the female wasps fly out into the world in search of a new fig tree. The lottery of life starts all over again.

This tale illustrates how advanced the interactions between flowers and pollinators can be. Unbelievably enough, this strange system where two types of common fig trees are needed to produce crops has been known to us humans since ancient times, and people used to hang goat fig branches in edible fig trees to get the whole business to work. The fig tree is apparently one of the earliest types of tree to have been systematically cultivated by humans.

Nowadays, we sometimes use cultivars that produce edible figs entirely without pollination and wasp visits. In any case, though, you needn't worry about finding a dead wasp when you bite into a fig. The goat fig the larvae are able to live in is hard and inedible, and in the edible fig variety, the trapped female wasp is broken down by enzymes and vanishes.

There are more than 800 species of fig tree worldwide and all typically have their own fig wasp to take care of pollination. These intricate partnerships have existed for millions of years: there are fossils of fig wasps with fig pollen that are 34 million years old, although the relationship is assumed to go at least twice as far back in time as that.

We humans aren't the only ones who like figs either. As a species, fig trees are reckoned to be the most important fruit trees of all in the tropics. At least a tenth of all bird species and a sixth of all mammals eat figs. And if this super-team of figs and fig wasps gets a little help with the move, they can even contribute to other services, like helping to rebuild vanished forests.

Krakatau is the name of a volcanic island group between Java and Sumatra, notorious for some tremendous eruptions over the past few centuries. One of these occurred in

1883, when great swathes of the largest island were literally blown up, making what is said to have been the most powerful noise in recorded history. The eruption wiped out all life on the island, but with a little help from fruit-eating birds and old-world fruit bats on the neighbouring islands, fig wasps and fig seeds found their way back there, and fig trees have since colonised the barren lava island. Today, 20 or so fig tree species grow there, providing the foundation for an ever-growing entourage of accompanying species. This has inspired scientists to experiment with the use of fig trees to restore dwindling forest areas in other tropical regions – with considerable success.

CHAPTER 4

A Well-stocked Pharmacy

An armour-plated sea creature, a brightly coloured North American lizard, an evergreen tree in an ancient forest ... What do these different species have in common? They have all provided us with medicines that have saved millions of lives.

Over the centuries, we have coaxed the willow tree into revealing its fever-suppressant secrets, which gave us aspirin. From the poppy we obtained morphine. The beautiful foxglove is the source of the digitalis medicine group, used for centuries as a heart medicine. In the Amazon, indigenous groups used to mix an arrow poison, curare, from different plants. After curare became known in Europe, it was used as a muscle relaxant in Western medicine and is still used in the early stages of anaesthesia to this day.

Every year, around 1 trillion dollars' worth of pharmaceuticals are sold worldwide. Even in today's hi-tech, synthetic world, more than a third of these medicines still come directly or indirectly from species in nature. For some types, such as antibiotics and cancer medicines, the proportion is much higher: between 60 and 80 per cent have a natural starting point. Raw material for new medicine still awaits discovery in hundreds of thousands of plants, fungi and animals.

Wormwood Versus Malaria

Over the millennia the plant kingdom has been our richest source of active medicinal ingredients. The oldest known documentation of medicinal plants is a 5,000-year-old clay tablet from Sumer containing recipes for 12 different medicines. More than 250 plants are involved, including several we now know to contain substances that act on the central nervous system, such as mandrake, henbane and poppy.

Many traditional medicinal plants gained their status after centuries of trial and error. That is why ethnobotany, the study of humanity's relationship to and use of plants, can be a good place to seek out new medicines. In 2015, Tu Youyou won the Nobel Prize in Medicine for her discovery of the active ingredient artemisinin, for use in the treatment of malaria.

Her finding was a result of decades of targeted searching based on traditional Chinese medicine, a project in which more than 2,000 Chinese herbal medicines were investigated in search of potential active ingredients that would be effective against the malaria parasite. In the end, the research group was left with a pale-green bushy plant with barely visible flowers: sweet wormwood, a close relative of mugwort, the bane of all pollen allergy sufferers, and of common wormwood, the plant used to flavour the cult drink absinthe.

The Chinese scientists realised that the plant contained interesting active ingredients but struggled to isolate them. In the 1,700-year-old *Handbook of Prescriptions for Emergencies* written by a Chinese herbal physician in the third century CE, Tu and her research group found important tips on how to extract the active substances from the plant.

Tests showed that artemisinin quickly and effectively kills the most dangerous malaria parasite (*Plasmodium falciparum*), with few side effects. This was good news, for in many places, malaria parasites have become resistant to previous remedies. Now treatment involves artemisinin combined with another kind of malaria medicine to make it more difficult for the malaria parasite to develop resistance.

Demand for artemisinin is high so scientists have been seeking a way to produce it in the laboratory. And this is where our friend brewer's yeast – Oregon's state microbe (*see Chapter 2, page 16*) – comes into the picture again. Since 2013, a genetically manipulated brewer's yeast has been busily producing raw material for malaria medication in great vats at a pharmaceutical company. Scientists are simultaneously working further to identify new and cheaper ways of producing the active substance and ensuring that the treatment is available to those who need it most.

Artemisinin is said to have been the most important break-through in efforts to combat malaria for several hundred years and more than a million human lives have been saved by this humble plant. In other words, it makes sense to safeguard traditional knowledge about species with cura-tive properties and follow up on it with scientific studies. This is crucial if we are to distinguish between supersti-tions with no scientific basis – such as the use of rhino horn – and knowledge that can lead to the development of new medicines. Nowadays, this kind of traditional knowl-edge is dying out in many parts of the world, as modern lifestyles and general urbanisation take hold and as indig-enous peoples abandon their traditional ways of life.

At the same time, the right to use and patent natural resources is a cause of conflict: respect for local knowledge clashes with the demand for profit, often accompanied by unpleasant baggage from colonial times. There have been countless cases of what is known as biopiracy – where foreign pharmaceuticals companies have exploited the knowledge of locals to earn massive amounts of money without the originators in the indigenous groups or local communities seeing any benefits whatsoever.

This is not simple. After all, who really owns the rights to a plant or a frog? And how should copyright of and income from shared goods and services of nature be distributed? Nowadays, there is an international treaty designed to regulate this, called the Nagoya Protocol, but it is still a challenging issue.

A recent example involves another potential malaria medicine. A state-owned French research institute (Institut de Recherche pour le Développement, IRD) interviewed 117 representatives of indigenous peoples and other inhabitants of French Guyana in South America, in search of plants and animals used to combat malaria. The 34 remedies suggested included a plant in the bitterwood family. This is a tropical plant family whose members also include the tree of heaven, which is frequently used as an ornamental tree in cities worldwide because it will happily grow along streets where air pollution is high. (Incidentally, male and female flowers grow on separate trees, as with willow, and people avoid planting the male trees because their flowers have an unpleasant odour.) The plant used to treat malaria is a smallish tree with beautiful red flowers known as *Quassia amara*. And I'm afraid I can't resist yet another brief digression: the Swedish naturalist Carl Linnaeus (1707–78) named this plant after Graman Quassi – a freed African slave and healer from French Guyana's neighbour, Surinam, who was already using the plant to treat fever in the 1700s. This was combined with the word *amara*, meaning bitter in Latin, because the medically interesting grazing deterrents in the leaves make them bitter.

And indeed, it was in the leaves of *Quassia amara* that the French scientists found a new substance to combat malaria parasites, called simalikalactone E (SkE). The IRD sought and obtained a patent for the substance in 2015, but failed to involve the authorities in French Guyana. Only after massive accusations of biopiracy from several quarters did they change tack and agree to share any profits with French Guyana, which was after all the source of both the knowledge and the plant material used in the study.

A Messenger Bearing Medicinal Mushrooms

Several years ago, I spent hours in a queue on a sun-baked street in a northern Italian city. I was keen to see Ötzi the Iceman – that poor soul who ended his days in an Alpine glacier 5,000 years ago. Like a wizened, hollow-cheeked messenger from Europe's Copper Age, he lies in his climate-controlled case as if belatedly laid in state. Few humans have ever been so thoroughly analysed as him: he's been X-rayed and CT-scanned and tested and studied in every conceivable way.

But what most interested me about Ötzi was what he had *on* him and *with* him. Because the cool, dimly lit museum also had display cases containing the clothes and tools found with the man. As an insect enthusiast, I was interested to see the remains of deer flies and two fleas that were found in his hair and clothes. Ötzi was also carrying various species of fungi, including some processed tinder fungus he kept in a little leather pouch. It may have been used to light fires or stem bleeding, like a primitive bandage.

He also had two round clumps of birch polypore fungus threaded on a string. One theory is that they had a religious, symbolic meaning. Another is that they were used as medicine to treat intestinal worms. For, of course, somebody has also analysed the unfortunate mummy's gut contents and found whipworm, an intestinal parasite. Although the worm medicine theory is controversial, there is certainly a long tradition of using birch polypore in folk medicine – to hinder bacterial growth, for example. A new study confirms that birch polypore contains active ingredients that may have medicinal potential, and asks whether the fungus could become a source for both medicine and biotechnology. We'll probably learn the answer in another 5,000 years' time.

It's hardly surprising that fungi can have medicinal potential, by the way. Fungi species are immensely numerous and many have strange lifestyles – living in soil, in hard-to-digest wood, inside living organisms. Consequently, they may have developed unique adaptations and can offer solutions to the challenges of life.

A classic example of medically significant fungi is, of course, the species of the *Penicillium* genus. This genus was the origin of penicillin, the first antibiotic ever discovered. Penicillin is deemed to be the most important medical breakthrough of the past 100 years – one of the fungal kingdom's greatest gifts to humanity.

Another frequently mentioned example is the immuno-suppressant medication ciclosporin, which is essential for organ transplants. This comes from a fungus that lives in the soil on my country's very own Hardanger Plateau, 'a bleak treeless plateau in the South of Norway' as a rather old Saudi Arabian article about the medicine gloomily puts it. The Swiss firm that collected the soil sample now earns billions of kroner a year from medications based on the fungus found on our dismal, desolate Norwegian plateau.

Yew's Whispered Wisdom

This is the time of tension between dying and birth
The place of solitude where three dreams cross
Between blue rocks
But when the voices shaken from the yew-tree drift away
Let the other yew be shaken and reply

T.S. ELIOT

'Ash Wednesday'

The yew tree has saved many lives through the cancer medicine taxol, named after the tree's Latin genus name, *Taxus*. The yew also has a long and complex history in human myths and literature, as well as its practical use.

First, let's zoom back 400,000 years in time to the Essex coast of England, two hours east of London by car, where the town of Clacton-on-Sea now stands. In those days – one of the interglacial periods of the Pleistocene – this was a fertile river plain, part open, part wooded and dominated by deciduous trees. People lived here. They weren't our species, although it is unclear which of our close relatives they were. What is certain is that they were surrounded by a rich diversity of large animals that we, sadly, will never get to meet. Archaeologists have uncovered bone fragments from steppe mammoths, straight-tusked elephants, giant deer, a couple of rhinoceros species, wild horses, aurochs and steppe bison.

This is also where the world's oldest wooden tool comes from: the sharp end of a spear. And it is made of yew.

Yew wood has special properties. It is strong, yet flexible. Long after most of the animals from Clacton-on-Sea had died out and vanished from Europe, yew was still being used in weapons. Ötzi, the 5,000-year-old iceman, was carrying an unfinished bow made of yew, as well as a copper axe with a yew handle.

Powerful yew longbows determined the outcome of several battles, undoubtedly influencing European history, especially that of England. At the Battle of Agincourt on 25 October 1415, during the Hundred Years' War between England and France, the English archers' arrows rained down on the much-larger French army with deadly efficiency, winning the battle. Historians believe that day was, at the time, the bloodiest battle in human history.

In Norway too yew was being used: in the archipelago of Hordaland in the west it was used to kill minke whales that were captured in the arms of the fjord until as late as the 1900s.

The modern application of yew, as a cancer treatment, started in the 1960s, when the US National Cancer Institute collaborated with the Department of Agriculture in a huge effort to find new cancer medicines in nature. Over a 20-year period, they collected and screened more than 30,000 species.

One hot August day in 1962, one of the botanists on the project found himself in a forest conservation area in Washington State beneath a shabby, 8-metre-high Pacific yew. Since the tree was the 1,645th plant from which he had gathered samples, it was assigned the plain and simple name of B-1645. The samples were sent for analysis and B-1645's bark was found to contain paclitaxel, a substance that stops cancer cells from dividing. Nonetheless, the path to patients was long and complex. Only in 1990 was the yew-based active substance approved for use against ovarian and breast cancer, and other types thereafter. To date, it is among the most economically profitable cancer medicines ever produced. The global paclitaxel market yielded earnings of almost $80 million in 2017, and the figure is expected to double towards 2050, driven by rising demand. And it all started with a scrap of bark.

But success can be a double-edged sword. Pacific yew was previously viewed as a worthless tree, almost a weed of the forest. You might think that the discovery of this significant active ingredient would have changed matters, leading to better protection for this important tree. The problem is that the tree needs to be stripped in order to extract paclitaxel from its bark. Although that can be done to living, standing yews (which will then die), in practice the trees were chopped down. A lot of trees. It takes 10 tonnes of yew bark to isolate 1 kilogram of taxol. That's equivalent to

3,000 trees – but still only covers a tiny fraction of demand on the world market. Add to this the fact that Pacific yews are only found in natural old-growth forests on the north-west coast of the US, are dispersed and few in number, and are among the slowest-growing trees in the world – and you can see the problem.

At the same time as taxol was becoming increasingly well known and more widely used, protests against logging on the West Coast of the US and Canada were also growing, because these forests were home to a rich, unique ecosystem containing many species other than Pacific yew. It became clear that new methods of producing the substance would have to be found. First of all, in the 1990s, a technique was developed that produced paclitaxel from the needles of the European yew, a different and more common species. Later, the pharmaceuticals industry developed purely lab-based techniques for producing the active ingredient. But both the Pacific yew and several Asiatic yew species remain on the International Union for Conservation of Nature (IUCN) global red list.

In Norway, European yew grows in a broad belt along the southern coast from Østlandet to Agder and up through western Norway. It is not on the global red list but is considered vulnerable in Norway and is therefore on the national red list. Molde, a municipality in western Norway, is famous for roses and jazz and the Royal Birch – the tree

King Haakon and Crown Prince Olav were pictured under when they sought shelter from German bombs in April 1940. But the municipality can also boast of being home to another tree – the world's northernmost wild-growing yew. It is often the case that marginal populations, meaning individuals that grow on the outer edges of the geographical range of their species, can have particularly interesting genetic properties. That is an additional argument for taking better care of the species in Norway. If you should be lucky enough to encounter one of the scattered, low-growing yews along the Norwegian coast, it's worth bearing in mind that everything on the tree, except for the beautiful red arils or seed coverings, is poisonous – both to humans and to many animals. There's a hair-thin line between poison and medicine.

Perhaps precisely because of its poisonous nature, yew has been seen as the tree of death. Ancient dark evergreen yews are a regular feature in cemeteries. And in literature, the three Weird Sisters in Shakespeare's *Macbeth* added yew to their deadly witches' brew. But the tree can also signify life and rebirth, and represents the transition between life and death. In pre-Christian times, the Celts considered the yew a sacred tree and believed it could carry the voices of the dead into our world as a whisper. It is this whispering that T.S. Eliot plays on in the lines from his poem 'Ash Wednesday': *This is the time of tension between dying and* birth/The place of solitude where three dreams cross/Between blue rocks/But when the voices shaken from the yew-tree drift away/Let the other yew be shaken and reply'.

The yew tree can also live a very long time – one yew in a churchyard in Fortingall, Scotland, is estimated to be somewhere between 2,000 and 3,000 years old – and branches that lie close to the ground can put down roots and form new trunks. There is a great deal of symbolism in the fact that the active ingredients from precisely this tree species are said to be one of our best plant-based cancer treatments. Through paclitaxel, the yew genus has given many people new life. Who knows how many more secrets the yews of the world will whisper in our ears if only we take care of them?

Monster Spit Slays Diabetes

I once suffered through one hour and 14 minutes of a terrible black-and-white horror film from 1959. In *The Giant Gila Monster*, a low-budget B-movie, an angry overgrown lizard wreaks havoc in small-town Texas (it was, incidentally, filmed back-to-back with *The Killer Shrews*, which wasn't exactly Oscar material either). The director drafted in a French former Miss Universe contestant in floral 1950s frocks to beautify the set but failed to film the correct lizard. The unintentionally comic scenes in which a genuine lizard moves through a miniature landscape star a Mexican beaded lizard, not a Gila monster.

The Gila monster is North America's largest lizard, at roughly half a metre in length. It is covered in bead-like scales in a beautiful psychedelic pattern in orange and black that looks like a batik project gone awry. In addition, it is one of the few known lizards equipped with venom – which it delivers by chewing on its prey so that the venom from the saliva glands in its lower jaw penetrate it. The species lives in the semi-desert southwest of the US, primarily in Arizona, as well as further south in Mexico. Poaching, building developments and road construction are causing their numbers to decrease steadily and the IUCN now considers the lizard to be 'near threatened' with extinction.

The Gila monster has a dark past as a misunderstood and unpopular creature. For a long time, people believed it had poisonous breath, that it killed its prey by breathing on it and that its bite was fatal to humans. None of this is true. That said, close contact with the creature can be extremely painful – both the bite itself and the effects of the venom, which is like 'hot lava coursing through your veins,' according to a YouTube celeb who makes his living showing off on-camera with things that bite and sting (there wasn't much showing off going on when the lizard unexpectedly took a chunk out of his finger, mind you!).

Among other effects, the venom acts on the pancreas, where insulin is produced. Insulin regulates blood sugar in the body and if you have Type 2 diabetes, the pancreas either produces too little insulin or doesn't work properly. This connection piqued the curiosity of a researcher and diabetes doctor in the Nineties. Financed by a modest basic research grant from the US government, and blissfully ignorant of the sensational finding that awaited him, he took a closer look at the Gila monster's poisonous saliva. What he found was a substance called Exendin-4, which resembles a hormone in humans. Exendin-4 boosted insulin production when blood sugar levels were high – for example, straight after a meal – and that helped keep the blood sugar steady at the correct level – just what diabetes patients need.

The scientist made a poster in which he described his findings and took it with him to the American Diabetes Association's annual conference. Here, he caught the eye of a small biotech firm and some 10 years later, in 2005, the medication was approved for use in the US. It is used as supplementary treatment by many patients with Type 2 diabetes, and since one of its advantages is that it has a lasting effect, it doesn't have to be injected so often. It also suppresses appetite, so it can help improve weight control in diabetes patients. In 2017, more than 1.5 million prescriptions were issued for the medication in the US alone. Fortunately for the threatened lizard, the active ingredient is easy to produce in the laboratory so there is no need to use living lizards to manufacture all this medicine.

That said, it's a good thing from our point of view that the Gila monster is managing to hang on in there out in the desert, squeezed in between all the motorways and construction projects. Because it turns out that its spit also has other interesting properties, like a capacity to influence memory. Laboratory mice have abruptly acquired super-memories – a bit like the plot of the Daniel Keyes novel *Flowers for Algernon*, in which first a mouse and then a person undergo experimental treatment that boosts their intelligence. We're not quite there yet in the real world but several pharmaceuticals companies are currently investigating whether substances extracted from the lizard's saliva can be used to treat memory loss in patients with Alzheimer's, Parkinson's, schizophrenia or ADHD. A summary article from 2019 discusses the use of variants of the active ingredient in Gila monster saliva to treat severe progressive diseases of the central nervous system. Although the article notes that a lot more human testing is needed, the interim results from animal tests are promising. The Gila monster has definitely been upgraded from movie villain to medical superstar.

Blue Blood Saves Lives

You probably don't know this but if you've ever had an injection, you owe a debt of gratitude to a sea creature with baby-blue blood and the looks of a medium-sized frying pan, which is responsible for ensuring that the contents of the syringe were pure and free from harmful bacterial poisons. Meet the horseshoe crab, a distant, ocean-based relative of the spider, which has saved masses of human lives in the past 25 years because its blood reveals whether or not bacteria are present in places where we don't want them.

Horseshoe crabs – known in Norwegian as 'dagger-tails' – have lived on Earth since long before the dinosaurs and have looked roughly the way they do now for the past 400 million years. They spend most of their lives in the sea but during mating season thousands of them simultaneously crawl up onto the beaches. Of the four species alive today, one lives on the East Coast of the US and the other three in Asia. The body of a fully-grown horseshoe crab is covered in curved armour plating with a thin, pointy armoured tail sticking out at the end. Although it looks a bit like a dagger, it's not a defensive weapon but more like a rudder to help the creature steer its course when swimming or walking. It also uses its tail to turn itself over if it happens to flip onto its back when on land.

The horseshoe crab breathes with book gills, large flaps that resemble the pages of a book, and oxygen is transported around its body in blood that contains copper. These copper compounds are what give the blood its characteristic pale-blue colour. Ten eyes are neatly distributed across the top of the horseshoe crab, and there are 10 feet on its underside. These allow the creature to shuffle its way through the mud of mangrove forests and shallow seas, as well as helping it to shovel food – various types of worms and mussels – into its mouth.

Characteristic horseshoe crab footprints dating back several hundred million years have been found in China, perfectly preserved in stone. Horseshoe crabs were among the few survivors of the 'Great Dying', the Earth's third mass extinction event 252 million years ago, which wiped out 96 per cent of all species in the ocean. The cause was a massive volcanic eruption in Siberia, which led to dramatic changes in temperature, pH and the oxygen content of the sea. But horseshoe crabs survived. Scientists can literally follow in their footsteps to see how horseshoe crabs stubbornly crawled their way through the mass death.

Fast forward a few hundred million years to our own times. Picture a lab where white-coated workers in hairnets and facemasks work efficiently side by side at long benches. Above the benches are rows of mounted horseshoe crabs. Their hinged rear sections and 'daggers' are tucked beneath

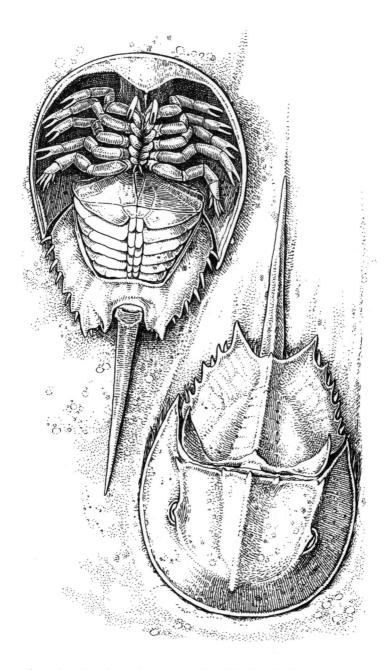

The ancient horseshoe crab has strange, baby-blue blood, which we use for producing safe vaccines.

their bodies to give easier access to the tissue around their hearts. From there, a thin cannula runs down into a glass bottle, which slowly fills up with pale-blue liquid, the horseshoe crab's bizarre blood. It may look like a scene from a sci-fi movie (remember Luke drinking blue milk for breakfast in the 1979 *Star Wars* film, *A New Hope*?), but this is a horseshoe crab blood bank, in which we humans play the role of bloodsuckers.

Humans first discovered the unique properties of horseshoe crab blood in the 1950s, thanks to two curious American scientists who were following up on some unexpected observations. One of them, studying blood circulation in horseshoe crabs, had noticed that the blood sometimes formed a jelly-like mass. He brought another professor on board who was working with bacterial poisons and their effects on blood and bleeding.

Eventually, the two scientists realised that the blood immediately clotted if it came into contact with bacteria. Even a tiny dose of endotoxins – a common bacterial poison from living and dead bacteria that can cause fever and, at worst, death in humans – was enough to make the horseshoe crab blood take on a jelly-like consistency.

Since sterilisation cannot destroy these types of bacterial poisons, it is vital to have a way of detecting them. Horseshoe crab blood turned out to be exceptionally good for this. Equipped with a bit of blood from this living fossil, it now became possible to test whether medications and medical equipment were safe to use. In 1977, the method was approved by the American health authorities and adopted worldwide. The clotting agent from the horseshoe crab blood is used to test all sorts of implants, injectable medicines, and vaccines – including COVID-19 vaccines – for the presence of harmful bacterial contaminants. And this is big business: a single litre of ready-to-use horseshoe crab blood is worth around $15,000.

Before horseshoe crab blood came onto the market, injectable preparations had to be tested on rabbits in a process that not only took more time but was also less reliable, so the new discovery saved thousands of rabbit lives. That said, life became tough for horseshoe crabs in both North America and Asia.

Every year, half a million American specimens and an unknown number in Asia are collected to have their blood drained in the 'blood bank'. In the US the process is regulated: quotas have been established, only a third of the blood is drained and the animals must be returned to the sea where they were captured within 72 hours. Even so, independent studies show that the mortality rate is around 15 per cent, and a significant number of American horseshoe crabs are also harvested for use as bait. The American horseshoe crab is listed as vulnerable (VU) on the database of the world's threatened species.

In Asia, the capture of horseshoe crabs is unregulated and the situation is apparently even worse there. The creatures are rarely returned to the sea after the bloodletting but are often used as food instead. In addition, the beaches where they gather are undergoing development, with the construction of houses and hotels. As a result, all the Asian species are on the global red list: one species is endangered (EN) and we lack sufficient information for the two others to place them in the correct category (DD or data deficient).

The decline in horseshoe crabs also has a dramatic impact on other species in coastal ecosystems. During the creatures' romantic assignations on the beaches, millions of blue-green caper-sized eggs are laid in the sand. In North America, this 'horseshoe crab caviar' provides a crucial energy boost for several species of migratory birds en route from South America to the Arctic. If you are a red knot who started your spring flight in Tierra del Fuego in Argentina, you'll definitely be peckish by the time you reach Delaware. But in recent years, the population of this subspecies of red knot has plummeted to less than a quarter of its 1980 level. One reason for this is that less food is available at the stopover spots, such as the horseshoe crab beaches in Delaware. Other factors include construction, increased disturbance, rising sea levels and climate change.

The demand for horseshoe crab blood has set this ancient creature on the road to extinction – in the same way as the discovery of active ingredients in artemisinin from wormwood, taxol from Pacific yews, exendin from the saliva of the Gila monster (and the vanilla flavour from the vanilla orchid) caused demand for these plants or creatures to sky rocket. Today, we no longer rely on the plants or creatures themselves to obtain all these substances, though: we can replicate them in laboratories, not just with chemical processes but also through biotechnology.

We humans have been practitioners of simple 'kitchen table' biotechnology for thousands of years – using brewer's yeast to ferment grain into alcohol, say, or lactic acid bacteria to acidify milk for yoghurt. But the really big breakthrough came in the 1970s when we learned to cut and paste genes, using methods known as recombinant DNA technique or gene splicing. This allowed us to insert foreign DNA into a cell and 'reprogram' a bacterium or a yeast, for example, to produce a particular protein for us. In recent years, new methods – in particular, something known as CRISPR technology – have made genome editing more applicable, simpler and cheaper. The biotechnology revolution offers new solutions in medicine and health. It brings many new challenges into the bargain, both professional and, above all, ethical: what are the risks and where are the boundaries?

It is undoubtedly positive for horseshoe crabs that we can now produce the enzyme that reacts to the presence of bacterial poisons in a cell culture in a lab without using the creature itself. It has taken a long time to introduce the new testing method but following its approval at the European level in late 2019 (applicable from 2021), there is some hope that this alternative can at last obviate the need to harvest horseshoe crabs and tap their veins. Let's just hope that we're not too late and that horseshoe crabs will continue to survive for another few million years.

Drugs from Bugs – Insects as a New Source of Antibiotics

Are you bald and keen to get your hair back? Try rubbing your scalp with a mush of crushed houseflies. Do you have urinary tract problems? Find seven furniture beetles in a dead deciduous tree, boil them up in milk and chug it down. History is full of peculiar advice about the ways different insects can help us with health problems. Amid all the superstitions and curiosities, some of the advice may also contain a grain of truth. Take this tip for toothache from *Die Tiere in der Deutschen Volksmedizin Alter und Neuer Zeit*, published in 1900: 'Those with toothache will enjoy relief if they help place a great many beetles that are lying on their backs on their feet again.' It sounds crazy and often enough the advice has no effect, yet it is true that certain beetles *do* secrete substances that have a mild analgesic effect. Take leaf beetles that graze on willow trees.

Willow contains an active ingredient called acetylsalicylic acid (ASA), familiar to us from medicines like Disprin, Aspro Clear and Caprin. If you turn over 'a great many of these' beetles and simultaneously touch your sore tooth in such a way that the substance enters your system, you may actually get some pain relief.

The fact is that insects may prove to be a treasure chest of active medicinal ingredients in future. There are many reasons for believing this. For one thing, insects are an extremely numerous group, with an estimated 5 million-plus species. They can also be found pretty much anywhere, other than the sea, and engage in countless complex inter-actions with other species – like eating the leaves of the willow tree and becoming a kind of six-legged headache tablet. Another kind of collaboration is even more impor-tant: we know that several insect groups engage in advanced chemical symbiosis with bacteria; these small partners produce anti-bacterial substances as a defence against other microorganisms that cause disease; it's not so very different from our own reliance on antibiotics. Take the ants in South America who farm special types of fungi. These ants keep other, harmful fungi away by providing fungicidal bacteria with specially designed 'bedsits' in a hollow in their bodies.

Another example is the beewolf, which is a kind of wasp, more specifically a digger wasp. At first glance it looks like a regular yellow-and-black-striped stinging wasp but it is larger and lays its wings flat on its back when it is not flying. What's more, the beewolf is not content to feed its larvae with little dead flies, the way stinging wasps do. No, what it needs is honeybees, and they have to be alive.

The beewolf gathers honeybees, paralyses them and then takes them into ingeniously designed tunnels in the sand, placing them in chambers at the end of the passageways. Between three and six bees are organised into a neat pile. It's a bit like you laying the breakfast table with cornflakes and juice for the kids before you head off to work because beside the piled-up breakfast the female beewolf lays an egg, which will become a larva. Which will eat its way through the provisions its mother has supplied for it. The female beewolf sets up several of these nurseries with ready-stocked larders in her tunnel system. She flies out and in with new bees until each of the nurseries is equipped with its own larder.

The final stage of the female beewolf's caring duties is to paint the nursery ceiling. The 'paint' is a white gloop carried in special glands in her antennae, which she squeezes out, like toothpaste from a tube. Perhaps the white roof is an emergency exit sign to help her young to get their bearings and find their way out once they are fully developed. But the paint has several functions and is superpaint – life-saving paint.

It turns out that this white substance from the mother's antenna is pretty much boiling with bacteria – 'good', co-operative bacteria of the *Streptomyces* genus. When the larva is all full up and ready to pupate, it incorporates these bacteria from the ceiling paint into its pupa case. This is no bad idea if you're about to lie around in a damp cave with all kinds of fungi and filth in the surrounding soil for an entire autumn, winter and spring as you wait for the following summer to come. It makes sense to have a pal like *Streptomyces* woven into your sleeping bag then, because the bacteria produce a whole tiny cocktail of different antibiotic substances. It's not so different from the combined treatments we humans sometimes use to prevent bacteria from developing resistance.

One of our most serious global health problems is antibiotic resistance – where pathogenic microorganisms develop the ability to withstand antibiotics owing to indiscriminate use. According to a 2019 study, 33,000 people die every year in Europe as a result of resistant bacteria. Another study estimates that by 2050, we can expect more people to die of antibiotic resistance than cancer: 10 million people every year, 14 times more than today. If the current trend continues, we risk seeing our grandchildren die of the same diseases that killed our great-grandparents.

Around half of all the antibiotics we humans use stem from the *Streptomyces* bacteria genus. Nowadays, it looks as if there is little more fresh material to be gleaned from the soil-living bacteria of this genus. And that's where insects come in, because this same genus of useful bacteria also exists in abundance on and in ants and wasps, beetles, flies, butterflies, moths and other bugs.

When a team of researchers recently checked more than 1,000 different insect species in search of new active ingredients to combat 24 bacteria and fungi that make us humans ill, they found that microbes carried by insects were much better than soil-based bacteria at fighting off antibiotic-resistant microorganisms. Lab tests on one of the new antibiotic substances – extracted from a Brazilian fungus-farming ant – indicated that it may work, on mice at least.

As always in the pharmaceuticals industry, the road to obtaining a finished medicine is long and winding, but insects offer a promising lead in the hunt for new antibiotics. Maybe that'll help us resign ourselves to the fact that we won't get a luxuriant new mane of hair from rubbing crushed houseflies into our scalps.

When the Kids Make You Puke

Once upon a time there was a frog. A seemingly ordinary greyish-brown slimy creature, caught in a stream in an Australian rainforest in 1973. True enough, the scientist thought it might be a new species but that wasn't so extraordinary in itself in a country with 240 species of amphibians (in the UK, there are a meagre seven). The frog, which was a female with an unusually big midriff, was given a spot in the lab aquarium. Nineteen days after her capture in the forest she was about to be moved into a new aquarium when, to the astonishment of the scientist, she abruptly regurgitated and spat out six tadpoles, and followed up a few days later by literally throwing up some fully formed baby frogs.

Suddenly this frog was far from ordinary. Instead, she became the world's only known species capable of swallowing her fertilised eggs and using her stomach as a uterus – hence the highly appropriate name assigned to her: the gastric-brooding frog. The frog young live in the mother's belly for six or seven weeks, developing from egg via tadpole to baby frog. Once they are ready, they are vomited up over a period of several days. During this time, the mother does not eat and nor does she produce gastric acids – if she did, her young would die inside her. In other words, what we have here is a species that can switch its gastric juice production on and off and can repurpose an organ when needed. This was of interest to medical science: what if we could find substances that would also regulate gastric acid production in humans? Or work out how to reprogram one organ to fulfil the function of another?

Other scientists captured more female gastric-brooding frogs to document this remarkable phenomenon and arranged a photo shoot in the lab. As the froggy mum was lifted out of the aquarium for some family snaps, her belly muscles contracted and a series of baby frogs abruptly shot out in what the accompanying scientific article soberly refers to as 'projectile vomiting' – landing some 60cm away. Other baby frogs turned back at the exit – the scientists saw them peeping out of their mum's open jaw before turning on their heels and promptly getting swallowed up again. Maybe they just didn't like what they saw, or perhaps they got an inkling of what lay in store for their species.

If the latter, it's hardly surprising they refused to hop out. These early articles about the gastric-brooding frog, all in the present tense, make for sad reading nowadays: 'The gastric-brooding frog *is* an aquatic frog ...' and 'the gastric-brooding frog *is found* only in a limited area ...' But the gastric-brooding frog no longer *is*, nor can it *be found*. Despite intensive search, nobody has succeeded in finding a single individual since 1981 and the IUCN has declared it extinct. Ironically enough, just a few years later a closely related species was discovered in the same area – also a gastric-brooding frog – but it too is now extinct. This means that medical science has lost the opportunity to study the mechanisms of the gastric-brooding frog and we will never know what medical discoveries it could have given us. Considering the life-saving medications we have found in far more ordinary organisms – like the common wormwood in China or the Gila monster in the United States – this is a sad thought indeed.

Nobody can say why the gastric-brooding frog vanished. Maybe it was the logging along the stream where it lived, or invasive species like weeds and domestic pigs that have escaped into the wild, or the fungal disease Bd, which threatens many of the world's amphibians. In all, well over 40 per cent of all amphibians are at risk of extinction according to the Nature Panel review – almost three times the level for birds (13 per cent).

Some scientists are keen to reverse the extinction of this little Australian oddball. They have extracted the gastric-brooding frog's genetic material from frozen frogs' legs (and other body parts) abandoned in some laboratory freezer and hope, by implanting it into eggs from a closely related frog species, to be able to bring the frog back from the dead. So far, though, the appropriately named Lazarus Project has got no further than producing a cluster of cells. While there's no denying that this is interesting, I'm among those who believe we should safeguard still-living species and their habitats first before investing large amounts of money in these kinds of de-extinction projects. Maybe you won't win generous research grants or prestigious awards for dedicating your time to good old-fashioned habitat protection, but you'll probably save a lot more species.

Mini-jellyfish and the Mysteries of Immortality

The gastric-brooding frog should have taken a leaf or two out of the book of hydrozoans – tiny, fragile relatives of the better-known moon and stinging jellyfish species – because some of them can almost live forever. The *Turritopsis dohrnii* species, often called 'the immortal jellyfish', can run its life on repeat – a fact of great interest to medical science.

Like most other similar hydrozoans, *Turritopsis* starts out in life as a tiny, free-floating larva called a planula. Eventually, the planula attaches itself to the seabed and starts to grow into a polyp, which initially resembles a little bush but ends up looking like a pile of plates. When the time is ripe, these 'plates' loosen and become medusae, umbrella-shaped forms that float away. So far so normal. But instead of behaving the way medusae are meant to – growing up, becoming adults, reproducing and dying – *Turritopsis* can jump back to the polyp stage and have another go. And another. All they have to do is not get eaten because that brings immortality to an abrupt halt. It's like a chicken abandoning the idea of growing into an adult hen and opting to go back to being an egg instead.

And just as no one believed scientists when they said they'd found a frog that gestated tadpoles in its belly, no one initially believed the scientists when they told them about the immortal jellyfish either. Because this kind of thing simply shouldn't be possible. At the start of an organism's life, all its cells are the same – so-called stem cells – but gradually, as the individual grows, the cells specialise and cannot go back to being stem cells again. Except that, in the case of *Turritopsis*, this is precisely what happens.

Scientists believe that this peculiar hydrozoan can teach us more about how to control cells, about how we can get the body to repair damaged tissue. The most optimistic – and original – scientist working with *Turritopsis* is an elderly Japanese researcher, Shin Kubota, the only person in the world who has managed to keep the species alive in laboratory conditions for a prolonged period. He believes *Turritopsis dohrnii* may hold the answer to the mystery of immortality, and is working to actively promote this little lump of jelly – he's even made YouTube videos of songs in its honour.

Just as insects are under-researched on land, marine invertebrates are under-researched in the ocean. Many scientists believe this is where we will find the medicine of the future. Over the past 50 years, more than 30,000 potential new active medicinal ingredients have been isolated from marine species, giving rise in turn to more than 300 patents. In 2019 news broke that after a year of targeted searching, a research group in Tromsø had found a totally new molecule in another small jellyfish (*Thuiaria breitfussi*) that kills the cells of an aggressive form of breast cancer. Marine bacteria and sea fungi (of which there are hundreds living in ocean environments) are other exciting groups worth investigating.

In the oldest great work of world literature, *The Epic of Gilgamesh*, the source of human immortality is said to be a thorny plant that grows on the ocean bed. Although it's true that *Turritopsis* is an animal rather than a plant, and it is far from certain that immortality is either possible or desirable for humans, the tales from these 3,000-year-old clay tablets nonetheless have a point: the ocean undoubtedly contains species with the potential to improve and prolong life.

Securing the Foundations of Nature's Pharmacy

A pangolin is a cat-sized mammal covered in big brown scales – like a kind of animated pine cone. With its almost unfeasibly long tongue – which is rooted in its pelvis, not its mouth – it scoops ants and termites out of towers and hidey-holes that it first rips open with its long claws. Pangolins are otherwise peaceable creatures that are mostly active at night. They don't even have teeth but chew their diet of termites with keratinous spines in their stomachs.

But it wasn't the pangolin's anatomical peculiarities that abruptly grabbed the attention of the world's newsrooms in spring 2020. Nor was it the fact that this unique mammal is on the brink of extinction – all eight Asiatic and African species are threatened and are on the global red list, as well as the CITES international database of species that are illegal to trade. No, the pangolin's celebrity status stemmed from the suspicion that it might have played a role in the transmission of the coronavirus from bats to humans.

One might ask how a threatened creature with a population in free fall could come into close enough contact with humans to serve as a potential vector of infection. The answer lies in superstition. In China, the pangolin's shell was once used to make armour, but what makes pangolins saleable today is the stubborn misconception that their shells possess medical properties. What's more, pangolin meat is used in exclusive luxury dishes. As a result, it has been possible to find both living and dead pangolins at outdoor markets in Asia, and this is how people believe coronavirus transmission may have taken place.

We can only hope that the resulting attention will give this mammal a slightly better chance of continued existence. The corona crisis has, at any rate, directed a spotlight

A pangolin is a cat-sized mammal covered in big brown scales – like a kind of animated pine cone. Illegal trading of pangolins might have played a role in the transmission of the coronavirus from bats to humans.

on the challenges of the wet markets where living creatures are sold, from the point of view of both human health and animal welfare. In addition, the pangolin was removed from China's official list of approved medicines in June 2020.

While the pangolin has the highly dubious honour of being the world's most illegally traded creature, it is far from being alone in these statistics. Illegal trade in vulnerable and rare species, often for use in traditional medicine but also as pets, is a billion-dollar industry. Animals, animal parts, timber and plant products from threatened species lead the pack when it comes to illegally traded goods, along with drugs and weapons. This trade is as lucrative as the drugs industry but with much less risk of getting caught. Internet trading and mobile phones have made illegal trade even easier to organise and harder to detect, and the scope just keeps growing. The burgeoning middle classes in Asian countries are especially keen buyers, but Europe also plays a significant role in this game, not least as the intermediary link in many of the transactions.

In June 2019, Interpol and the World Customs Organization organised a joint operation to combat this illegal trade. In 26 days, they confiscated, among others, several rhinoceros horns, hundreds of kilograms of ivory, 23 living primates, more than 4,000 birds (many living and stuffed in bottles with their beaks taped shut), nearly 10,000 living turtles and 1,500 living reptiles. The mortality rate of wild-caught reptiles for illegal trade is, incidentally, so high that it is comparable to that of cut flowers.

The 30 big cats that were confiscated included a white tiger cub that was hidden in a cage in a truck in Mexico – perhaps en route to the US, where the number of tigers in private ownership apparently exceeds the combined global population of wild tigers. In this case, it's not a matter of medicine but status. Believe it or not, estimates suggest that between 2,000 and 5,000 tigers are being kept, like miserable overgrown alley cats, in Texas alone (the Netflix hit series *Tiger King: Murder, Mayhem and Madness*, among the most watched on streaming services in spring 2020, shows the dismal and unethical conditions many of these creatures live in). There are now said to be fewer than 4,000 wild tigers in the entire world.

Over-harvesting and illegal trade are just two of the threats to our planet's diversity. They come on top of dwindling nature and destroyed habitats, climate change, the movement of species and various types of pollution. There is no doubt that we could learn a great deal more about the active medicinal ingredients to be found in nature, yet we continue to put these crucial species at risk for reasons more aligned with personal gain than with the greater good. We are also on the threshold of a new era, in which we can use ecological knowledge to find new bioactive substances in nature, which we can then produce synthetically in laboratories. This gives us the opportunity to make a medicine while sparing wild populations. But for that to happen, we must realise that we have to do a better job of conserving the starting point for new discoveries: species diversity. Today, according to estimates, we are losing at least one important new medicine every other year as a result of our disrespectful dealings with the planet's natural pharmacy.

CHAPTER 5

The Fibre Factory

The pharmaceuticals industry isn't the only one scouring nature for resources and active ingredients; the industry and technology sectors are doing the same thing too. In your everyday life you are surrounded on all sides by different types of fibre produced by plants and trees. The clothes you wear, the walls of your home, your shelves, books and all, the firewood that warms your living room on a winter's day – all this comes from nature's fibre factory. But there are other, less obvious, applications for natural fibres too: as flavouring in vanilla ice cream or food for farmed salmon. There is a lot to be gleaned, especially from interactions between trees and fungi, where the processes that occur shed light – both literally and figuratively – on new ways of using natural fibre.

From Fluffy Seed to Favourite Fabric

Have you ever heard of mallow? In case you're wondering, it has nothing to do with the marshmallows you toast around the campfire. No, it's the name of a plant family. And there's a good chance that you're wearing at least one item of clothing made with fibre from this family as you read this. It's a fibre that runs through our history in an unbroken thread – from the oldest find in an 8,000-year-old grave in Pakistan, via the seeds of the Industrial Revolution to today's environmentally damaging textile industry. I am referring, of course, to cotton, the world's most widely used textile fibre.

We do not know where cotton was first grown but we do know that different species of the cotton plant were in use in different parts of the world, independently of one another – including the Mehrgarh civilisation, in today's Pakistan. Mehrgarh is among the oldest archaeological sites in southern Asia, and traces of agriculture and animal husbandry found there date all the way back to 7000 BCE.

Human graves at this site have taught us various things. This, for example, is where we find the world's earliest examples of the art of dentistry: nine poor wretches – four women, two men and three individuals of unidentified sex – bear clear signs of having had their teeth drilled, apparently with some kind of flint tool. In another, earlier grave lie an adult man and a child aged about two. Around his left wrist, the man is wearing a string of eight beads made of copper. Analysis showed that each of these copper beads contained remnants of cotton fibre, the remains of the string that joined them; this is the oldest example known to date of humans using cotton.

Almost as fascinating is the discovery of well-preserved 6,000-year-old scraps of woven cotton fabric in a blue-and-white pattern from a totally different part of the world, Peru. Here, people used cotton for fishing nets and textiles. The blue colour of the cotton yarn apparently came from the dye of the indigo plant.

Sheep's wool was well known in Europe but cotton had to be imported. And perhaps because so few medieval Europeans had seen a cotton plant, a peculiar myth sprang up of a hairy sheep plant that was said to grow in remote regions of Asia. One variant of the myth describes it as a kind of tetherball hybrid of plant and sheep: a real live sheep was rumoured to sit atop a sturdy stalk – complete with flesh and blood and wool. This stalk, a combination of stem and umbilical cord, was flexible enough to allow the sheep to graze on the ground below for as far as the stalk stretched. The myth was only dispelled in around the 1700s.

Cotton also plays a role in more recent history. Consider the part it played in the plantation slavery of the United

States, or England's Industrial Revolution, which started with the use of spinning machines such as the Spinning Jenny in the production of cotton textile.

When you pull on a pair of jeans, you're essentially putting on trousers made of dried fruit. Cotton fibre is those long white hairs that grow out of the seeds of the cotton plant. Each hair is a single long cell, and one seed can produce 10,000–20,000 such seed hairs. Many kinds of plants have hair or down on their seeds – think of cotton grass or the dandelion's familiar fluffy head – but cotton's properties are unique. Of all the plant hair in the world, cotton hairs are the only ones with the combination of length, strength and three-dimensional structure in a dry state that makes it possible to spin them into thread or yarn. That said, we also make textiles from plant stalks and leaves, such as flax, hemp or bamboo.

Cotton fibre can absorb up to 25 times its weight in water and becomes stronger when wet. That is why we don't just use it in clothes but also in bandages, ecological sanitary pads, towels – and banknotes. Our Norwegian notes, with their images of cod and Viking ships, are printed on cotton paper. The Bank of Norway believes this makes it easier to incorporate security features.

Nowadays, half of all textiles produced contain cotton. Over the past 30 years, the cotton-growing area has remained constant at around 2.3 per cent of the world's agricultural land. In the same period, production has almost doubled, owing to the use of more intensive cultivation methods. But cotton production is not especially environmentally friendly, as it uses a lot of water, fertiliser and pesticides. Cotton is a thirsty plant: it takes at least 10,000 litres of water to produce a kilogram of cotton fabric, enough to make a pair of jeans and a T-shirt.

Looking at insecticide sales alone, a full 14 per cent of the total is used for cotton cultivation (2009 figure). What's more, most of the cotton grown today is genetically modified – a process whose impact is disputed. At the same time, the vast scale of cotton-growing ensures work and income for many people: more than 250 million are involved in the cotton sector, plus many more if you include the textiles industry. The challenge now is to switch to more environmentally friendly processes of production, using less water and toxic chemicals. Because after 8,000 years of use, cotton is unlikely to go out of fashion any time soon, even if the Bank of England has started to replace its cotton pounds with plastic notes.

Home Sweet Home

In summer 2019, I was one of the teachers on a course about biodiversity in dead wood. A collaboration between Norwegian and Russian universities, it took place at

Voronezh Nature Reserve, close to the town of the same name in southwest Russia. During a fascinating journey involving Russian night trains and encounters with many delightful people, there were two high points for me: getting to see male stag beetles – those huge beetles whose tremendous jaws are almost longer than their bodies – on full display on the trunk of an old oak and visiting a local museum of early human history.

The Kostenki Museum is far from imposing on the outside: a square block of a building, it looks like a Norwegian construction supplies warehouse that has been dropped from the heavens and landed in the middle of the Russian village, among tiny ramshackle houses and modest plots of land. But don't be deceived by appearances. The museum is constructed over another building, one that has been here for 20,000 years. In those days, the region was a steppe, held tight in the grip of permafrost. There was barely a tree to be seen back then. So, what did the humans of the day use to build their dwellings? Mammoth bones, of course.

At the museum in Kostenki, I gazed straight down upon the excavated remains of a circular house, or perhaps rather a tent, made of bones. Various parts of the mammoth skeleton were stacked up in a framework, reinforced by whatever wood was available and clad in reindeer hide to form a kind of lavvo (tent). The area around Kostenki (whose name means 'bone' in Ukrainian) is among the earliest modern human sites so far found in Europe. Mammoth hunters were already living here 45,000 years ago and the region is packed with the bones of mammoths and humans alike.

Building materials in the form of bones, hide and wood are examples of tangible goods and services of nature. In Norway, we have been building with wood since the pithouses of the Stone Age – sunken dwellings with frameworks of wood and turf. There are many traces of such Stone Age dwellings in Norway and they are better preserved there than anywhere else in Northern Europe. A PhD thesis from 2017 shows that some of these houses were maintained and reused for up to 1,000 years.

Later, more advanced techniques took over, with a load-bearing construction of wooden pillars – as in the Bronze and Iron Age longhouses and the 1,000 or so stave churches we believe were built in Norway (of which 28 remain today). In the Middle Ages, log-building techniques became fully established. And Norwegians continued to build houses from timber, in the towns that later grew up too, until as late as 1904. However, wooden houses have one major disadvantage: they burn. At a quarter to two on the night of Saturday 23 January 1904, a fire alarm was sounded at Ålesund's canning factory. Fifteen hours later, 10,000 people had been made homeless by a fire so vast that it prompted a flood of assistance from people all over Europe – from the German Kaiser to the actress Sarah Bernhardt.

The fire in Ålesund led to a new law, 'Murtvangloven', which required houses in the centre of Norwegian towns to be built of stone.

Given our current focus on the environment and with new building techniques increasing fire resistance, wood is undergoing a renaissance as a construction material in Norway and the rest of the world. Knowing that concrete is responsible for 8 per cent of the world's CO_2, looking for more sustainable construction material has been given high priority. In 2010, Japan introduced a new law that made the use of wood mandatory in all new public buildings with fewer than three floors. But there is no reason to be so modest when it comes to the number of storeys: high-rises can also be made of wood. One of the world's highest wooden buildings, a 67-metre temple in China, has been standing since 1056 and has survived several major earthquakes.

Now so-called Plyscrapers have become the latest hot trend in architecture. For the time being, Norway boasts the world's highest wooden building: the Mjøs Tower in Brumunddal, eastern Norway, completed in 2019, stands 18 storeys and 84.5 metres high. Who knows how long it will hold the record since several other countries are mulling similar buildings. But the original is still the winner: the world's tallest tree, a coast redwood in California, measures 115.86 metres in its stockinged feet.

By the Light of a Fungus Lamp

Darling, I am writing to you tonight by
the light of five mushrooms.

AMERICAN WAR CORRESPONDENT IN NEW
GUINEA IN A LETTER TO HIS WIFE

From time to time, I am lucky enough to participate in
Abels tårn (Tower of Abel, named after the famous
Norwegian mathematician Niels Henrik Abel), a radio
programme produced by the Norwegian national broad-
caster. It's a popular science show where listeners send in
questions for us to answer. In spring 2019, a question came
in from a 95-year-old listener living in Kristiansund on the
western coast of Norway. He was 16 years old when the
Second World War broke out – and that was when he saw
something he has never been able to forget. During the
German invasion of Norway in 1940, Kristiansund suffered
massive incendiary bombing and was reduced to a smoking
ruin. Like so many of the city's other inhabitants, our man
was evacuated to one of the neighbouring islands in
Nordmøre, Tustna. It was there, as bombers roared over-
head, that he experienced this phenomenon he had been
pondering for the intervening 79 years.

This is the story he told: he had taken a walk into the
village and returned to the farm late. While putting away

his bike in the dark cart-shed, he saw a strange light coming from a woodpile in the corner: 'From the floor up to a height of around 30 centimetres, the wood was luminous. In my recollection, the light had a bluish sheen. I seem to recall that it appeared to be alive, as if it was flickering in a way. The light was so beautiful, and it seemed so mysterious that the bottom 30 centimetres were luminous while the rest of the wood was normal.'

I believe this must have been a phenomenon known as phosphorescent wood and that a fungus was the source of the peculiar light. Because although fungi cannot photo-synthesise, some species occasionally do the opposite: decompose organic material, such as wood, producing carbon dioxide – and light. The light comes from an energy-carrying molecule called luciferin. Its name derives from the Latin *lucem ferre*, light-bearer – although we now tend to associate the word with Lucifer, the devil, the fallen angel of light. When luciferin meets an enzyme called luciferase, light energy is released.

We know of around 75 species of luminous fungi, most of them in the tropics. The same bioluminescence occurs in many other organisms, including insects such as fireflies and glow-worms, and many marine creatures, particularly in the dark depths of the sea. In the ocean, various organ-isms use light to communicate, attract prey or scare off enemies. The purpose is less clear in fungi. Some think the light plays a role in attracting insects that can help spread the spores of the fungus. Others believe the function is rather to scare off fungus-eaters. A third possibility is that the light is simply a by-product without any clear ecological function.

This light, produced in living organisms, is special in that it is cold light, which does not radiate heat – unlike burning wood or heated metal, such as the filament in an incandescent light bulb. It wouldn't make much sense to burn your own body, after all.

This kind of cold light was practical in the olden days when people wanted to light up places where a fire would be particularly undesirable so it is logical that old historical works from the Nordic region in the 1500s describe people using bundles of rotten, luminous oak bark to light their way up to the hayloft in the dark.

In war too phosphorescent wood proved useful. During the First World War, soldiers in the trenches fastened pieces of luminous rotten wood to their helmets at night-time to avoid bumping into each other in the dark. American soldiers did the same thing on night patrols in the jungles of Asia during the Second World War. Indeed, an American war correspondent posted to New Guinea even wrote home to his wife – by the light of five luminous mushrooms.

Other times, phosphorescent wood has been a nuisance. During the Second World War blackout in London, the glow of the timber yards along the Thames could be so bright that the men on fire-watch had to cover the luminous wood with tarpaulins.

One curious example of the use of phosphorescent wood dates back to the American War of Independence in the 1770s. The American inventor David Bushnell designed *Turtle*, the world's first submarine. The plan was to use it to attach explosives to an enemy ship, below the water line, in order to break the British blockade of the Port of Boston. The use of phosphorescent fungi in the submarine is described in a letter dated 1775: '*On the inside is fixed a Barometer, by which he can tell the depth he is under water; a Compass, by which he knows the course he steers. In the barometer and on the needles of the compass is fixed fox-fire, i.e. wood that gives light in the dark.*' Good old Bushnell deserves recognition for his creativity but unfortunately the phosphorescent wood didn't work very well. In a later letter he describes how the compass needle stopped glowing. Apparently, conditions were too dry or too cold for the fungus in the scrap of phosphorescent wood.

One type of fungus we know can glow is honey fungus or *Armillaria*, a genus that is widespread across the entire northern hemisphere (this genus, incidentally, includes what may be the world's largest individual organism: a subterranean honey fungus in Oregon covering an area of roughly 10 square kilometres). Honey fungus can exist as a parasite on living trees, forming black, metres-long strings beneath the bark that look like flattened liquorice shoelaces. The process that creates the light is apparently concentrated in the tip of these black shoelace strings.

At Norwegian latitudes this is probably the fungus behind the phosphorescent wood. Honey fungus was also the answer I gave to our listener, who had spent a lifetime wondering what it was that lit up the cart-shed that April night in 1940. I'm inclined to believe that the wood in the woodpile was fresh and damp, and thoroughly overgrown with the threads and strings of the honey fungus. A dark woodshed and a light-carrying fungus molecule gave one teenager an unforgettable glimpse of nature's very own magic.

Chanterelle's Clever Cousins

I've been on the lookout for luminous honey fungus myself for several years, to no avail. In the meantime, I settle for edible fungus, also a source of intense experiences. An autumn mushrooming trip in beautiful forestland can make me forget absolutely everything else. I'm like a woman possessed – I'll go just that little bit further, then a tiny bit further still ... all in the hope of spotting a yellowish-orange cluster of chanterelles shining at me. My mushroom mania is restricted to a minuscule part of this vast and complex kingdom. It is mostly focused on chanterelles, hedgehog mushrooms and porcini, but part of the fun with forest hikes in autumn is also looking for and philosophising about all the other fungi to be found.

Fungi are wildly fascinating organisms – to my mind, they come a close second to insects. Many people might think that the fungus kingdom is closely related to the plant kingdom, and perhaps that's hardly surprising, since the two groups can resemble each other in appearance. The chanterelle you pick in the forest is the fungus's reproductive organ, just as the flower is the plant's. The bulk of the fungus body consists of a network of mycelia or fungal threads, which looks identical to the root system of a plant, superficially at least. Both fungi and plants live static lives and, like plants, fungi consist of many relatively similar modules combined into an organism (whereas animals' bodies are generally composed of non-repeating, specialised organs). Together these modules create a large surface and that is useful, whether an organism needs to photosynthesise or get enough food – without having to chase it.

However, if you overlook the superficial, it becomes clear that fungi have more in common with animals. As one of the professors from my student days liked to say: fungi are actually inside-out animals. They don't fiddle around with photosynthesis but – like animals – rely on breaking down biomass built up by plants. Instead of internal digestion in the form of a gut system, though, fungi practise external digestion – their stomachs are on the outside. From the surface of their bodies, fungi give off digestive enzymes that dissolve the surrounding biomass. The nutrients are then absorbed through the cell walls and into the fungal body. New methods that use DNA to examine kinship have also managed to establish that fungi are closer to the animal than the plant kingdom. Perhaps good old Linnaeus, the natural scientist who classified and named so many species, was more correct than he guessed when he placed fungi among the animals in a genus he called *Chaos*.

Fungi also share another property with my beloved insects beyond fascinating lifestyles: they are a diverse, species-rich group. In Norway alone, we have mapped 8,418 different fungal species and calculate that there are almost half as many again. You'll never see hide nor hair of most of them in the forest but that doesn't mean they don't matter – quite the contrary. Some fungi, in the congenial company of insects and bacteria, help break down all the dead matter in the great outdoors. It's a crucial job that enables nitrogen and carbon recycling. Other fungi are sponges and parasites; they don't care whether their host is dead but will happily attack living plants or animals as well. We aren't as fond of all those types. Athlete's foot, for example. A third group is mycorrhizal fungi. (This peculiar word – which you're guaranteed to spell wrong the first time you write it – comes from the Greek words *myko*, meaning fungus, and *rhiza*, meaning root.) Like the fungus world's answer to the extrovert, social individual, they live in an eternal coexistence with plant roots, looking after nature's underground internet. Through their very own Wood Wide Web, trees and plants can exchange nutrients and chemical substances in something akin to a form of communication.

The first group especially, the decomposers, contains numerous fungus species that have proven useful to industry. Allow me to introduce *Obba rivulosa* – one of the species I think of as the chanterelle's clever cousins. It is a white-rot fungus that grows in dead trees, both conifers and deciduous species, often in trees partly charred by forest fires. The fruiting body of the fungus forms on the surface of the wood in diffuse yellowish-white clusters with distinct pores. It can look like dollops of dried foam. No great beauty, perhaps – though if beauty comes from within, this species definitely has its charm.

It turns out our *Obba* has a somewhat unusual appetite and contains enzymes that can selectively break down some of the most inedible components of a log (in case you've forgotten your science lessons: enzymes are biological substances that accelerate various reactions in animals and plants). What's more, *Obba*'s enzymes can do the job perfectly well, even at low temperatures. This is good news, as it reduces energy use and waste in paper production. In 2003, Finnish scientists submitted a patent application for the industrial use of *Obba*'s enzymes.

Obba rivulosa is not found in Norway. Another white-rot fungus called *Funalia trogii* is, however – a beige-white, scruffy, hairy specimen of wood-decay fungus. It survives by breaking down dead aspens and here too special enzymes are involved, a type known as laccase. Inspired by the fungal world and *Funalia trogii*, industry is now making liberal use of laccase – to clean coloured and poisonous waste water, to break down unwanted by-products in oil refineries and to bleach paper pulp.

But *Funalia trogii* may have even more tricks up its sleeve. In laboratory conditions, an extract from the fungus has proved capable of killing cancer cells without harming normal cells. The cancers concerned are largely hormone-dependent variants, such as breast cancer, ovarian cancer and testicular cancer. And it is *Funalia trogii*'s variant of the super-enzyme laccase that gets the job done, although researchers haven't quite worked out why yet. Other variants of laccase, from a handful of other wood-decay fungi, are also in the spotlight.

I must swiftly point out here that many potentially promising active medicinal or biochemical ingredients fail to make it through the gruelling process of gaining approval for medical use or prove too unprofitable to be scaled up for commercial use in industry. However, the point is that there are unexploited resources out in the forest that can do more than fill my basket with chanterelles – a lot more.

Campfire Contemplation

There is so much to forests. The trees alone offer enormous possibilities. We have already discussed building material. Another obvious product is firewood.

The conquest of fire and the ability to extract energy from dried, dead organic materials like twigs, grass or animal dung are among the earliest and most fundamental of humankind's ingenious discoveries. Campfires helped us by providing light and warmth, warding off wild animals and giving us a new diet of roasted and boiled food. Eventually we also used fire deliberately to shape our land-scape – to improve grazing or clear fields for crops.

Energy from fibre remains important to this day. Globally, more than 2 billion people depend on wood for energy. In Norway, around 8 per cent of the energy we used in 2018 was derived from various types of biofuel. A good third of that came from firewood, the rest from pellets, wood chips and liquid biofuel.

By the way, do you know how you can check whether your birch firewood is dry or not? Spread a thin layer of washing-up liquid and water on one end of the log and blow into the opposite end – and see if you can blow soap bubbles. Birch wood contains microscopic open-ended wooden tubes stacked on top of one another, so it's like a long drinking straw. While the tree is alive, these tubes transport water from the root up to the canopy. In fresh birch wood, residual water in the tubes will make it difficult to blow through them, but if the wood is dry, the air will go straight through the log, producing tiny soap bubbles at the opposite end.

I'm fond of wood-fired heating. In the cabin where I spent time as a child, the woodstove in the bedroom had windows made of transparent mica. From where I lay in the bottom bunk of the single bunk bed, the gentle, flickering glow of the fire animated all the faces I thought I could spy in the beams and planks of the bunk above me. With these woody creatures for company, I drifted happily off to sleep in the dim light, lulled by the reassuring murmur of grown-up talk in the room next door.

Something still falls into place inside me when I gaze into the flames of a campfire. We humans are no longer very good at doing nothing. A campfire helps: sitting quietly and letting our thoughts off the leash like a pack of hounds in the mountains. Letting them bound away. Maybe they'll follow the path at first. But then they'll get bolder, changing direction and dashing away over heather and hillside, running off their restlessness until they suddenly come to a halt, nose buried in some exciting find. Before the camp-fire has burned to embers, our thoughts return home, calm now, and settle quietly at our feet.

Gazing into a campfire forges you into an unbroken chain of forefathers and foremothers who have done precisely the same over hundreds of thousands of years. It also hooks you up to nature's everyday yet fundamental process: photosynthesis. Because when you put a log on the fire and it flares up, what you are seeing and feeling is like delayed sunshine. Solar energy, carbon dioxide and water built up the biomass, the 'plant body' itself, that is – a tree trunk. Now you are releasing the sunbeams once again.

Sprucing Things Up: The Conifer that Flavours Food and Feeds Salmon

Imagine a freshly chopped Norwegian tree trunk. A good third of it is destined to become planks, chipboard and the like. The rest will become paper, pulp, fuel and numerous other biochemicals that will, in turn, be used in everything from toothpaste to concrete. It's truly incredible what you can conjure up from a spruce log, including products like vanilla flavouring and animal fodder.

Vanilla flavouring originally comes from vanilla beans – the seedpods (or, formally, seed capsules) of a beautiful pale-flowered orchid that grows in Mexico and further south. The Totonacs, an indigenous group on the eastern coast of Mexico, were apparently the first people to harvest vanilla. They have a legend about its origins: in mythical times Xanat, a divine being and daughter of the fertility goddess, walked among humans. She fell in love with a mortal man but since she was divine, they could not be together. Out of her mind with grief, she transformed herself into what we know as the vanilla flower, which still bears her name in the Totonac language: Xanat flower. This allowed her to remain on Earth and offer her beloved the joy of her incomparable beauty – while giving the rest of us the fragrance and flavour of vanilla.

In the 1500s, the Aztecs conquered Totonac territory and demanded tributes of vanilla beans. When the Spanish conquistador Cortés reached the Aztec capital, he was greeted with a vanilla-flavoured cacao drink. This is how Europeans first discovered the flavour.

The Totonacs continued to be the world's largest producers of true vanilla until the mid-1800s. Then – after years of fruitless efforts – the French finally managed to get vanilla plants to produce the valuable seed pods on the colonial island of Réunion in the Indian Ocean. The vanilla orchid isn't especially cooperative, you see. In

Mexico it is pollinated by a genus of stingless social bees (*Melipona*). These are not found on the islands of the Indian Ocean, and consequently, the vanilla plant remained unpollinated. A 12-year-old slave was the first person who managed to pollinate the plant by hand, with the aid of a stalk of grass.

But even after this technique had been refined, it took vigilance and patience to cultivate fully developed vanilla pods for sale. The daughter of the goddess visits the Earth only fleetingly – the vanilla flower withers after a day – and the seedpod's aftercare is complicated and takes months. In all, 40,000 pollinated vanilla flowers are needed to produce a kilogram of natural vanilla flavouring. It's hardly surprising that people were highly motivated to find simpler ways of creating the flavour.

At the end of the 1800s, the substance that gave vanilla its flavour, vanillin, was isolated, and its chemical structure determined. From there, it didn't take long to produce the substance synthetically, first from pine bark and later via a process that started with oil of cloves. A little less than 1 per cent of the world's vanillin is still produced using that method. A small but growing percentage is also made from rice, aided by fermentation of the rice husk. The majority, around 90 per cent of the world's vanillin, is made from oil, while a substantial share (around 7 per cent) is made from spruce trees – in Norway.

The vanilla flavour originates from the capsules of the vanilla orchid.
Today more than 99 per cent of demand for vanilla flavouring is covered
by other, synthetic sources.

Vanillin is simply a by-product of the paper industry and in 2020, a Norwegian firm in my birthplace of Sarpsborg, southeast Norway, is the only company in the world making vanillin in this way. It is working to increase capacity, because global demand for vanilla flavouring is increasing – and the original source, the vanilla pod, can only meet a third of a per cent (!) of world demand.

Tiny black dots in your ice cream are no guarantee that the *flavour* stems from vanilla pods. Even if the seeds are in fact genuine vanilla orchid seeds, they may be entirely flavourless – a waste product from the production of genuine vanilla extract. The seeds may be added to ice cream purely for visual effect, along with vanillin based on oil or wood shavings.

So, does vanilla from slivers of spruce taste exactly the same as vanilla beans? Vanillin in its pure form is, after all, precisely the same chemical substance regardless of its origin. But natural vanilla doesn't just consist of vanillin; it also contains tiny amounts of many hundreds of other substances that contribute to the taste experience. So, if you like making eggnog, it makes sense to use the real thing, to ensure that the complex flavours come to the fore. But if you're planning to bake, it probably isn't worth going to the expense of using vanilla pods. The combination of many other flavours and a long cooking time will cause the additional flavouring substances to disappear. That makes the lab-produced alternatives perfectly adequate substi-tutes. And that's a good thing. Considering that orchid vanilla outprices silver on a per-kilo basis and can't even cover 1 per cent of demand, it's handy that spruce trees can help spice up our lives with vanilla flavouring.

≳◦≲

Spruce trees are surprisingly useful. My university, the Norwegian University of Life Sciences (NMBU), is home to the Foods of Norway research centre, which does research into trees as a basis for animal feed. It uses sugar compounds from dead wood to grow yeast, which is then ground into high-protein flour. This allows salmon, piglets and chickens alike to be fed by the forest – that is to say, scientists have experimented with replacing a share of the proteins in ordinary fodder with yeast fodder.

Naturally, this raises the question of whether a wood salmon can be a good salmon. I may be fond of butter-fried chanterelles with salt and pepper, but can salmon be fooled into becoming fungus-eaters? After all, their natural diet is plankton, insects and small fish. So far, though, the trials look promising: the lab fish are growing well and have even shown improved intestinal health.

In the early days of fish farming, fishmeal was used as fodder but demand quickly outstripped the supply availa-ble from an overfished ocean. Currently, Norway's farmed salmon are fed on large amounts of soya imported from Brazil. That is hardly environmentally friendly – both because soya production may displace rainforest and because priority should be given to using soya for human food. In other words, if we're going to pursue large-scale fish farming to provide proteins for a growing human population, more sustainable fodder alternatives must be found – like spruce trees.

We aren't quite there yet, though. Further research is needed before this kind of animal fodder can be produced plentifully and cheaply. The road from concept to finished commercial product is like a hike in a rugged forest reserve: it's full of twists and turns and easy to get lost in, and all of a sudden, you're liable to come across a ravine that's well-nigh impossible to scale. This is what the innovation sector calls 'The Valley of Death': where good ideas go in, but nothing comes out. Even if something is technically possible, there's no guarantee that the game will be worth the candle.

Besides, although trees are a renewable resource and there are 3 billion of them in the world, a forest is more than just timber resources. Nowadays, ever more responsibility is being placed on the forest's green shoulders. It is supposed to purify water and air, rescue the climate, outcompete concrete and steel, replace soya imports to provide salmon feed, give us new products where oil is being phased out, safeguard threatened species, and provide us with mushrooms and berries and adventures in the great outdoors. And while the forest certainly gives us countless natural goods and services, it is equally certain that we cannot maximise all of them at the same time. When everybody wants a piece of the forest, we need to shift our gaze away from the trees and look at the entire ecosystem, from hedgehog mushrooms to humankind.

CHAPTER 6

The Caretaking Company

A t the office or at home in a housing co-op, we don't give much thought to the caretaker who goes to the trouble of adjusting the heat or stopping water from ending up in places where it's not meant to. That's how it works in nature too: like a true caretaker, nature mostly works behind the scenes. Trees and other vegetation retain water and soil. Forests on the steep mountainsides that are so abundant in Norway protect the roads and buildings alongside the fjords below from avalanches. Wetlands provide a buffer against floods, and coral reefs and mangrove forests protect against tsunamis. City trees muffle noise, clean the air, provide shade and regulate the temperature. This chapter is about nature's caretaking company, focusing especially on the places where we build and live.

Too Much, Too Fast, Too Polluted

As a 25-year-old, I travelled through the heart of Australia by bus to see Uluru (also known as Ayers Rock), a vast red island of a rock formation. Uluru was a truly imposing sight: higher than the Eiffel Tower and spread over an area of nearly 2 × 4 kilometres, the mountain lies out in the middle of the flat desert like a vast, half-buried beached whale. Tourists stream here to see the mountain's hue change with the angle of the sun, but you have to be extraordinarily lucky to experience what I did here: pouring rain. After all, it is a desert.

When the heavens opened their floodgates that day, I got to see the role vegetation plays in capturing rainfall. Since the hard, red sandstone surface of Uluru doesn't have so much as a blade of grass – it is, indeed, entirely free of soil – all the rain ran down its surface. The raindrops met, dripped, ran, gathered in hollows, came together in larger channels, grew into streams. Just instants after the downpour started, the rain came splashing down to the foot of the mountain where I stood, in cascades large and small.

This is precisely what happens in our cities: because we humans have eliminated vegetation and replaced it with surfaces that do not absorb water, the rainfall gathers on the tarmac and concrete, and quickly runs in vast quantities towards lower-lying parts of the city. Along the way, it gives buildings and infrastructure a battering. The river, where the surplus water generally ends up, may burst its banks, causing even more damage, because there's too much water for it to carry off.

What's more, the rainwater washes away all the chemical substances left on the tarmac by traffic and industry. It eats away at any soil it flows over, causing erosion. Both soil and pollution are washed out into the rivers. If the sun was baking the tarmac before the rain arrived, the water will also heat up during its passage through the city streets. As a result, it may be several degrees warmer than the water it joins in the river – and that can cause problems for the species in the river.

All in all, this is good for neither city nor river. And even though we can't control precipitation or prevent all floods, we can diminish their effects through better teamwork with nature – by making room for more greenery in the cities. Trees are pretty brilliant in cities as they have a few tricks for combatting floods: they are thirsty – a large tree can 'drink' several hundred litres of water a day. What's more, the canopy will serve as a brake, checking the rainfall before it hits the ground, or simply capturing it, so that it evaporates directly from leaves and branches. The tree's root system and the diversity of life in the earth around its roots make the soil more porous, enabling more of the rainfall to seep into the ground instead of flowing across its surface. Soil organisms associated with the tree can also absorb and convert harmful substances such as heavy metals.

We will probably have to live with large amounts of impervious surfaces in our cities, since planting vegetation on roads is no solution, and flower meadows make poor car parks. The more green spaces of all kinds we can retain in among the tarmac and concrete, though, the better it will be. One bonus of trees is that they can be planted alongside roads and pavements, creating a layer *above* the imperme-able surfaces. A meadow or lawn will also absorb and halt water, playing a similar caretaking role of keeping water away from places where it isn't welcome. And this kind of low-growing vegetation is also a suitable covering for the roofs of houses – green roofs.

This isn't a new concept: in Scandinavia, people have used turf on top of birch bark for roofing since prehistoric times. Indeed, turf roofs were also common in Norwegian towns throughout the Middle Ages: an engraving of Bergen from the late 1500s shows sheep and goats grazing on the roofs of many of the houses. As the city's wooden dwellings became more tightly packed together, turf roofs were banned, as dried grass on the roofs made it easier for fire to spread.

Now, green roofs are on the rise in cities worldwide. In some places, like Munich, it is obligatory to have vegetation on new buildings with flat roofs. A green roof has several advantages: not only does it capture rainfall, it also helps cool the city. And although you may not find flocks grazing on today's green roofs, perhaps at least you'll be able to take a stroll in the green, high above the city.

When Money Grows on Trees

City trees aren't just flood-catchers. A tree also serves as an air-cooler and air-cleaner, and provides living space for other species, including humans – you can climb it or sit leaning against its trunk reading a book in its cool shade.

Why is it always many degrees warmer in our cities than the surrounding areas? Several things come into play: in place of vegetation, which releases cooling moisture when it transpires, we have built dark surfaces of tarmac, masonry and concrete, which absorb heat when the sun shines. On top of that comes the heat from people and machinery, including the surplus heat from air-conditioning units, which, ironically enough, make it even hotter for people who don't have cooling systems themselves. As a result of all this, the major cities are like islands of heat, up to 5–10 degrees above the temperature of the surrounding land-scape. This temperature difference is often greatest at night, when stored heat is released from the dark surfaces of the city.

Now perhaps excessively hot summer days are no great problem for those of us at the top of the planet – for now. But climate change will make the planet burn beneath many people's feet. A Swiss study illustrates this by comparing 2050 climate forecasts for certain cities with the current climate in others: Oslo will become like today's Bratislava when the maximum temperature in the warmest month increases by 5.6°C, while London's climate will be like that of today's Barcelona – around 5.9° hotter at the highest temperature. If these forecasts materialise, trees will come in handy because they reduce the temperature in cities by between 1 and 5°C. This means saved lives. In Oslo, scientists at my second workplace, the Norwegian Institute for Nature Research (NINA), have taken a look at the significance of city trees and why we must protect them. They found that trees and other green areas effectively reduce the health risks caused by heat waves in cities: for every tree that is chopped down, one more elderly person on average is exposed to an additional day of temperatures above 30° over the course of a year. At the same time, the scientists point out that there is a huge need for more green areas in Oslo. The most highly populated districts are also the ones with fewest city trees.

Sometimes money actually does grow on trees, if only indirectly, because this cooling effect means we can save energy and therefore money. Studies from the US show that if tree canopy cover is increased by 25 per cent in a city like Sacramento, California, an average household will save 40–50 per cent of the energy they would otherwise have expended on air-cooling. And there's money in that.

Other savings will also result. Trees act as air-purifiers too, because tiny pollutant particles are captured and deposited on the leaves and branches. The next rainfall washes them down and into the soil or out into the river's waterways. Although the trees do not necessarily render the pollution harmless, they do at least remove it from the air we inhale. On a global basis, this is important. The Intergovernmental Science-Policy Platform on Biodiversity and Ecosystem Services (IPBES) estimates that only a tenth of the world's population breathes clean air. Every year, more than 3 million people die early as a result of air pollution – primarily in Asia. An ongoing study from more than 10,000 air-quality monitoring stations in a range of countries shows that the first two weeks of the coronavirus lockdown in spring 2020 alone led to a reduction in air pollution that resulted in 7,400 fewer early deaths.

It's not easy to slap a price label on all this, but there are systems for setting a value on trees in cities: London's most expensive tree, a plane, is supposed to be worth £1.6 million. In all, the city's 8 million or so trees offer total annual benefits of £132.7 million.

How Green Was My Valley –
Until the Topsoil Blew Away

There's a joke people tell about forests on Iceland. Short, but pretty much to the point. It goes like this: 'How do you find your way out of an Icelandic forest? You stand up'. The fact is that forests are neither plentiful nor tall in the land of the sagas, not any more. That leads to major challenges with erosion and land degradation.

A good 1,000 years ago, when Norwegian Vikings came sailing along, threw their high seat pillars overboard and settled wherever they washed ashore, there was a lot of forest on Iceland. Maybe as much as 40 per cent of the land was covered in trees – the same proportion of forest that we in Norway have today. But the settlers cleared away the birch forests in favour of cropland and grazing pastures. Trees were also needed for building material and charcoal. After a short time, 200–300 years, the country was pretty much treeless. Without trees to give shelter and hold the soil in place, Iceland's light, volcanic soil was laid bare – exposed and vulnerable to wind and rough weather. And there is a lot of rough weather in Iceland. And even more wind.

Erosion set in. Slowly but surely, the topsoil was blown into the sea, washed away or covered by sand drift. Volcanic eruptions, ash and high impact from grazing sheep added insult to injury. As the topsoil decreased, the vegetation dwindled even further, which led to even more soil loss. By around 1950, 60 per cent of the vegetation and a full 96 per cent of forest and bush cover was gone. Less than 1 per cent of the country was covered in forest. Iceland had become an open landscape. Fine for the tourists, perhaps, who get to enjoy unimpeded views of glaciers, volcanoes and mountains. And yes, it's photogenic in its brutal, barren beauty, which deploys the entire palette of earth tones, but it is not remotely fertile. In large parts of the country, soil loss and erosion have made crop-growing and grazing impossible.

Some years ago, I visited Iceland to take part in a conference with the European chapter of the Society for Ecological Restoration. In addition to hours of more or less exciting lectures, we went along on an inspection in south-west Iceland, to see what happens when we humans disable nature's erosion-reducing mechanisms, and what we can do to remedy it. The bus drove us for hours around a lava landscape, where patches of stubborn verdigris moss were the only vegetation in sight. We visited Gunnarsholt, a farm established before the 900s by no less than the grandfather of Gunnar of Hlíðarendi, the protagonist of part one of the thirteenth-century Icelandic saga called *Njál's Saga*. Gunnar dies in *Njál's Saga* – after he is declared an outlaw but cannot bring himself to leave his farm. He starts to ride off but reins in his horse, looks back at the beautiful arable landscape and says: 'Fair is the slope: fairer it seems than I have ever seen it before, with whitening grain and the home field mown: and I shall ride back home, and not go abroad at all.'

Gunnar probably didn't spare so much as a thought for erosion or the caretaking services of nature. It might have been a good idea if he had. Because several hundred years on, the slope is no longer so fair. Many farms, including his own, Gunnarsholt, have had to be abandoned owing to the ravages of erosion. Nowadays, Gunnarsholt is the headquarters of *Landgræðsla ríkisins*, Iceland's Soil Conservation Service, and a tiny, interesting museum. Here, we heard about Iceland's efforts to bring the forest and its sought-after services back to the island.

The last stop of the day was a flat plain covered in a pitted carpet of low vegetation: lupines. Along with all the rest of the diverse collection of scientists, I was handed a metre-long planting tube in blue metal, and two small birch plants. Eagerly, we spread out across the terrain: action at last – Iceland shall be reforested! The planting tube is placed on the earth as if it were a pogo stick and a hole is made in the soil by treading on a little plate at the bottom of the tube. Then I was able to drop my first ragged birch shoot with its five yellowish-green leaves into the tube, letting the tree sail straight down into its future home. All that was left to do was tread down the soil around the plant plug. My back was spared and Iceland gained another tree.

A symbolic action – because the work we did that day doesn't amount to much. Coaxing the forest back is slow work on a massive scale. First, the soil must be stabilised with plants like lyme grass or Alaskan lupines. The latter is a species introduced from North America. Although it is good at gaining a foothold where little or nothing is yet growing, and serves as a soil improver by binding nitrogen from the air, the widespread use of this invasive species is not uncontroversial. It spreads all too easily: in Norway, the Alaskan lupine is in the 'severe ecological risk' category on the alien species list – a dubious honour it shares with other undesirables such as Japanese knotweed, Spanish slugs and Canada geese.

When I asked the Icelanders about this, I got contradictory answers – people's opinions differed when it came to how heavily concern for native flora should weigh in the balance when the ecosystem was as messed up as it was here. The same discussion applies to the species of tree being planted, because birch – Iceland's principle native tree – isn't the only species shooting down the planting tubes. Large numbers of alien species such as larch, Sitka spruce, lodgepole pine (twisted pine) and poplar are also being planted or sown.

The government has a green dream: 12 per cent of Iceland will be forested by 2100. Today, the figure stands at around 2 per cent. It'll be a long time before you can get lost in a great Icelandic forest.

Flying Rivers in the Amazon

There is a river in the Amazon – a vast river that transports billions of tonnes of water. It's the basis for a fantastically rich diversity of humans and other species, and it influences climate and precipitation patterns over vast swathes of the South American continent. But it isn't the river you're thinking of.

The Amazon rainforest covers just 4 per cent of the Earth's surface, but is home to a tenth of all known land-based plant and animal species, as well as several hundred indigenous tribes. The trees in this forest outnumber the stars in the Milky Way. And because the trees are big, a tenth of the world's biomass is gathered here, in one huge, hot, green, muggy, pulsating ecosystem. And through this forest runs the Amazon in all its beauty and might: it is the world's largest watershed (equivalent in size to the US) and accounts for roughly a fifth of all the river water in the world.

Above the canopies hovers another, equally significant, river: a flying river formed of water vapour. It is created by the trees themselves and influences both rain and wind on a continental scale.

The theory of flying rivers and their consequences was first aired in 2007 by two Russian scientists, and its official name is the 'biotic pump theory'. Initially met with massive resistance, the theory has since gained a great deal of support. It goes like this: the whole process starts with the trees absorbing water from the soil and transporting it up to the greenery in their canopies. There it is used in biological processes, before evaporating and entering the air as water vapour – as if the tree were a geyser in slow motion: a geyser that sucks moisture from the ground and sends water vapour high into the air. When this water vapour condenses in the atmosphere, it creates low pressure, which draws more humid air inland from the ocean like an enormous flying river. According to the scientists, the trees send 20 billion tonnes of water up into this 'cloud service' above the Amazon rainforest every single day. That is, in fact, more water than the twin rivers on the ground dispatch into the Atlantic Ocean.

So, the rainforest acts as a living pump that transports moisture from the ocean into the continent's interior. This biotic, or living, forest pump with its prolonged low pressure draws rain to the Amazon and this, according to the scientists, is why there is more rainfall deep in the Amazon than on the coast – quite contrary to prevailing models.

141

When the flying river eventually reaches the Andes to the west, it curves away, continuing south. Along the way, it dumps the rest of its precious cargo of rain over regions south of the Amazon – some of the most important agricultural land in South America.

In order for this to work, it is crucial for the rainforest not to be logged and broken up: by doing that, we risk destroying the entire pump system. The results may be drastic. The most terrifying scenario is that we may reach a tipping point where we risk transforming the entire Amazon rainforest into a savannah-like landscape within a fairly short period. If that happens, the consequences for both humans and other species across the entire continent are enormous.

Termites and Drought

The climate is influenced by biology, and both large species, like the trees of the Amazon, and extremely small species, like insects, can be involved. If you could float above the semi-desert in northwest Tanzania in a hot-air balloon, you could – once you'd got over your vertigo – peer over the edge of the basket and gaze down on a strangely regular landscape. A sandy brown surface strewn with green dots, spaced at roughly equal distances. But even though it looks artificial, this is not a human-made landscape. The pattern is created by termites, small white or brown insects that look a bit like ants. In fact, termites are relatives of the cockroach that have learned to live together. Termites form advanced social colonies that move both water and nutrients, thereby shaping large parts of the world's hot, dry regions in Africa, South America and Australia.

But termites come in for a lot of criticism. There's even an entire industry dedicated to their eradication, which is hardly surprising when you consider that, in the US alone, they cause more than 2 billion dollars' worth of damage every year by greedily gobbling up beams, floors, walls and roofs. That's the way it goes when you belong to the tiny minority of animals on the planet that *are* able to digest the tough cell walls of woody plants. What's more, they can build up massive communities – a subterranean termite colony can consist of millions of individuals. If you were somehow able to weigh every single one of these tiddlers in, say, 10 square kilometres of African savannah, you'd find that their combined weight would match or even exceed that of all the large herbivores living in the same area.

But things that are a nuisance in the house are vital in nature. In the semi-desert and savannah environment, termites can be crucial organisms, contributing to both fertilisation and irrigation. They can gather dead plant residue, break it down and ensure that the nutrients get mixed into the soil in the form of dung or dead individuals, or passed on to the animals that eat them. Thus, in ecosystems vulnerable to fire, they prevent nutritious matter from simply going up in smoke. Termites can also dig deep down into the earth – as far as 50 metres – and bring up moist mineral soil particles that they use in their constructions. When this mineral soil weathers, it enriches the earth around the termite colony with critical nutrients, trace elements and moisture. The many tunnels the termites dig are also crucial to making the soil more porous, enabling rainwater to seep down more easily. All this makes termite towers tiny green oases in dry regions, and termites are key species that control much of the ecosystem. Their capacity to create these green 'hotspots' for plant life is important in the battle against desertification. These oases, which allow species to cling onto life, can serve as buffers against the desert. Plants can spread out from these oases in periods of precipitation. Studies show that these kinds of termite-tower landscapes are surprisingly robust in times of drought, meaning that termites can help stabilise the ecosystems in these sensitive regions in the face of imminent climate change.

The termites' role in maintaining a stable climate is not exclusive to savannah regions: in Borneo, scientists have studied termites' impact during periods of drought in rainforests by comparing areas where they eliminated virtually all termites and others where the colonies were left intact.

There proved to be a third more moisture in the soil where termites were present and could bring the moisture up from the deep layers of the soil. It's hardly surprising that the test plants the scientists put out had a 50 per cent greater chance of survival in the places where termites had not been eliminated. In the Asian rainforest too these much-maligned bugs turned out to be providing an ecosystem service that may be especially useful in a changed, drier climate. But in order for that to work, we have to take better care of intact nature and its native communities of species. Logging and other human incursions into forests alter the number and species composition of termites, and more than half of all tropical rainforest is already affected by humans. Tropical forests that have been drastically altered by humans are likely to be less capable of withstanding drought, according to the authors of the Borneo study, owing to a decline in the termite-driven buffer against drought.

There are almost 2,000 species of termite with different lifestyles. Some live in metres-tall, highly visible clay towers, others have underground nests or live in trees. Some eat dead wood, others feast on all kinds of dead plant matter and some grow fungus inside their towers. Sometimes these fungi create massive mushrooms that grow up from the tower – clocking in as the world's largest edible mushrooms, with a hat that reaches from your armpit to your fingertip.

A sought-after delicacy among the local population.

And speaking of large, termites in Brazil have built what is apparently the world's biggest continuous construction produced by a single species: a network of underground passages that connects metres-high piles of earth – 200 million towers in all. The whole thing covers an area the size of the UK and it is ancient, close to 4,000 years old – in other words, the same age as the pyramids at Giza. But unlike the pyramids, there is still life and movement in this 'building'. Strictly speaking, the towers here aren't nests and nobody lives in them. They are simply dumping places for all the soil dug out of the tunnels – the parts that are in use. The tunnels serve as covered walkways that make the trip to and from the termites' food in the nearest thicket quicker and safer.

If you don't have the opportunity to float above either Tanzania or Brazil in a hot-air balloon, there's still hope for you. Zoom in on Google Maps and take a look at the result of the termites' drought-controlling constructions from the comfort of your office chair – that's how monumental they are.

Mangroves as Breakwaters

The Vikings of Iceland and their descendants didn't see which way the wind was blowing until it was too late. Nowadays, we are well informed about the protective impact of forests, both in our northern forests and the tropics. Even so, short-term gains often hinder long-term and profitable conservation. One typical example of this is mangrove forests, which are vanishing at a lightning-fast rate, even though their significance as living breakwaters against tsunamis and floods is well known.

Mangrove is the name given to several species of trees that like to dabble their toes in saltwater. The term is also applied to the forests formed by these trees. Mangrove trees have adaptations that enable them to grow with salt waves lapping around their trunks and to gain a foothold in soft mud where there isn't much oxygen.

One such adaptation is their ability to grow stilt roots, multiple arm-thick roots that grow out of the lower part of the trunk – first, horizontally, then curving down into the mud. The net result is a tangle of supports that sometimes give the impression of a tree with legs, ready to set off at a run. In fact, the opposite is true: the stilt roots anchor the tree. At the same time, the intertwined roots serve as a buffer against waves, trap sediment and slowly but surely build up more sludge on the bottom for the tree to grow in.

If there isn't much oxygen in the mud, the stilt roots may have small bark pores that serve as ventilation channels but mangrove trees can also develop special breathing roots that grow in the opposite direction we expect roots to take: they shoot up from the mud into the air and look like little spears.

Life is hard for these forests, on the boundary between sea and land. If you study a world map showing the original distribution of mangrove forests, it looks as if somebody has taken a marker pen and drawn a line that follows the contours of the coastline from South Africa all the way to Japan, also outlining many islands, small and large, as well as parts of the west coast of Africa and both the east and western coast of Central America. This may sound impressive, but the fact is that the world's mangrove forests have shrunk dramatically – almost 40 per cent are already gone. Today, these forests account for less than half a per cent of all the forests in the world, and they have diminished at three to four times the rate of other forests.

As the mangroves vanish, so too does the work they perform in curbing the impact of extreme events along the coast. And where we try to imitate their effect with human-made constructions, it isn't just more expensive but less effective too. Take the super-cyclone that hit the Indian state of Odisha in 1999. It remains the most severe tropical cyclone in the North Indian Ocean to date, with a maximum wind speed of more than 300 kilometres per hour. Almost 10,000 people died, and the damage to homes and property amounted to more than $5 billion. But studies show that coastal villages with intact mangrove forests experienced lower fatalities and less damage to infrastructure than villages where the mangrove forest had been removed and replaced with a breakwater or dyke.

The mangroves lay there like a living, protective belt, cushioning the land against tidal waves, floodwater and wind speed on the leeward side. In addition, floodwater ran back to the sea more quickly, whereas the villages with human-created flood-protection structures struggled for longer with brackish water that destroyed their crops.

Equivalent studies were carried out for the tsunami in the Indian Ocean in 2004, with the same result: both loss of human life and material damage were lower where the mangrove forests were intact, and in the wake of the catastrophe the survivors' chances of economic development were far better in these areas.

And this is just one tiny aspect of the goods and services of nature offered by mangrove forests. The forest is also a unique and extremely rich ecosystem that purifies water, captures three to five times more carbon than other tropical forest and provides food, wood and other goods to 120 million people who live in and near it.

So why are they being destroyed and removed? Put simply: because you and I want to eat scampi. The main reason for the decline in mangrove forests is the opportunity for short-term aquaculture gains, especially among shrimp farmers. One challenge is that the amazing job the mangrove forest does in limiting damage from tsunamis, floods and hurricanes is not included in the calculations. Let's take a look at a concrete example, published in 2013.

An area of mangrove forest in Thailand can provide income through sustainable harvesting that conserves the forest. Alternatively, one can remove the forest and set up a shrimp farm. The latter option can give the owner a high individual income over the short term – especially large because generous subsidies have sometimes been made available for this kind of alteration in land use. But then we can start to calculate what this mangrove forest gives us in the way of natural goods and services, and what they are worth – cushioning the impact of waves, purifying water, providing a habitat for juvenile fish. That turns the calculation on its head: from society's perspective, there is very little economic gain to be had from sacrificing these precious natural goods and services in exchange for shrimp production. And if we add the costs associated with restoring a mangrove forest after a shrimp farm has polluted and destroyed the ecosystem, the difference in favour of the mangrove forest comes to just over GBP 16,000 per hectare.

This example illustrates how important it is for public efforts to be directed not just towards a single natural good or service – generally a product, like shrimp or timber – but to look at things holistically, to see the sum of natural goods and services being provided. Just as it is unprofitable to stimulate shrimp production in Thailand at the expense of the mangrove forest's protection against damage caused by nature, nor is it a good social investment to provide support for road building and timber operations on the steep mountainsides of western Norway if that destroys the natural, old-growth forest's ability to limit avalanches (and also destroys the rich species diversity found in such forests).

If you want to help conserve important tropical coastal ecosystems and the caretaking services they provide, choose a different seafood – or try and track down eco-certified scampi. Because although the mangrove forest only accounts for a few thousandths of the world's forests, its caretaker services and contribution to our welfare are far from modest. To chop down these forests and thereby lose the natural protection they provide along the coast is truly to saw through the branch we are sitting on.

Beauty in a Rotting Branch

In a valley
that none frequent
the mightiest tree
has fallen forward,
with wide outspread
branches and twigs
pressed to the soil
as if in an embrace
after endless yearning

(...)

and the tree will lie unmoving
deeper and deeper in its embrace
and begin to become the other –
while grass grows and falls over it
like pale, familiar tresses
– and all has long since passed
and a hundred years is a mere
instant
for what endures

TARJEI VESAAS

from '*Trøytt tre*' (*Tired Tree*)

I have a favourite fallen tree in the forest where I live. We are almost blood sisters, the tree and I, having exchanged a drop of blood for a spot of resin. Ten years or so ago, this old tree took it into its head to fall across the path I usually run along. It was a grey Sunday in October, one of those days when the sky has abandoned any idea of short, water-saving showers and it just pelts down endlessly for hours. I actually love running in the forest when it's raining but my glasses are a problem – and lenses don't work for me. I've opted either to see a bit with glasses – which fog up and stream with rainwater – or to see even less without them.

I'd brought my eldest out with me on my run; eight years old at the time, he was bringing up the rear. We were running along a steep hillside when I saw a little tree that had blown over, forming a low portal over the path. I ducked under it at full speed and ran onward, turning back to tell my boy about it. What I didn't see was that yet another tree had fallen across the path: a sturdy spruce, which slammed into my forehead when I turned back again, just that bit too late. I saw stars. All at once, there I was, prostrate among conifer needles and bark, my forehead sticky with blood and resin.

The result was concussion, and a two-week ban on screen-use and reading. Because unfortunately we humans don't have brains that dangle inside our skulls on an elastic band – unlike woodpeckers, which are thereby able to cope with bashing their heads against tree trunks.

Since this dramatic first date, I've followed the tree with special attention. First of all, the needles turned brown and slowly scattered, and bark beetles dropped by to make nurseries beneath the bark. Somebody, a keen pensioner perhaps, cut the tree and pulled the trunk aside so I no longer needed to duck beneath it on my runs. The following summer, the bark began to loosen here and there, because longhorn beetle larvae had eaten their way through the layer that holds bark and trunk together. I could see the pale wood dust created by thousands of chewing beetle young drizzle out as I ran past – a wood dust that was used to powder sore babies' bottoms in the 1700s, according to Linnaeus's reports from northern Sweden.

Soon the first red-belted conk appears, one of the most common wood-decay fungi found on spruce in Norway. At first, there's just a yellowish-white, shining clump, almost as if a lump of dough has been left inside the trunk to rise and is starting to ooze out. As the fungus grows, the red-belted conk's characteristic dark-brown and reddish-orange hues begin to emerge, often adorned with big, shiny drops of water. These fungal tears are not dew – i.e. condensed water vapour – but water that is squeezed out of the fungi in periods of rapid growth.

We humans tend to think of dead trees like this in the forest as mess, seeing the decomposition as something gloomy and unpleasant, redolent of decay and death. How very wrong we are. Because dead wood is alive. Inside 'my' fallen spruce there are now more living cells than when the tree stood green and sturdy. Where a living spruce bough would mostly consist of dead cells, there is now teeming, creeping, gnawing life. Wood-decay fungi extend their mycorrhizal threads through the cell structures. Slowly, the fungi's enzymes digest the structures that once held the tree up. In this way, the nutrients are made available to all manner of insects that eat their way through the growth rings. Add lichens, mosses and the odd terrified shrew that has sought shelter in the hollows of the fallen tree, and you'll understand why a third of the species in Norwegian forests live right here – on and in the dead trees.

I have studied forests and trees and their decomposers for a quarter of a century. It all started when, as a fledgling MA student, I was roaming about in the forest around the family cabin to gather 1,000 tinder fungi from dead birches. My plan was to see whether there were fewer beetles living in these fungi when they grew in isolation, on a birch that stood all alone (there were). In the intervening years, I have studied insects in all sorts of different trees and forests. I have gathered beetles from dry spruce and birch stumps on clear-cuts, from the foot of ancient pine trees and from aspens whose tops we lopped off at human height with dynamite to mimic natural standing dead trees. Almost every spring since I have been out to set up insect traps – in protected old-growth forests, charred forest fire sites, in the pillared halls of industrial forests. I have studied insects in the cavities at the base of linden trees that sprouted some time back in the mid-Holocene Warm Period maybe 5,000 years ago, where new trunks have grown up from the root system as the old ones died. The coarse roots resemble the arms of a massive octopus, snaking over the limestone along the Oslo Fjord. Over many years, I have followed insect communities in old hollow oaks – trees that have seen generation after generation of human life slip by as their buds burst, grow into leaves, then wither and fall, while the years, the decades and the centuries pass in the slow heartbeat of arboreal life.

Recently, we put together beetle data from all these years, composed of no fewer than 158,070 individuals from 1,267 different beetle species. We wanted to find out whether there were any overarching common traits that could explain why the species are where they are, whether it was possible to detect a signal strong enough to come through, even in such a mixed and variable dataset from more than 400 forest locales throughout the whole of southern Norway.

And we did find such signals. One factor emerged as significant, cutting across all variations: logging. Natural old-growth forests, which have never undergone clear-cutting and have not been planted, house a richer array of threatened or near-threatened beetle species than old managed forests. In all, most wood-living beetle species were found in the natural forest. The results from clear-cuts with dead trees lay in between. What's more, the composition of species was different in each of the three types; here, the age of the forest and the amount of forest (volume) within the landscape also played a role.

The results are not all that different from those we obtained in another study I was involved in, which dealt with wood-living fungi in dead spruce. In that study, we compared specialised wood-decay fungi with ordinary wood-decay fungi in 28 forests across the Scandinavian peninsula. The specialised fungi species, many of which are threatened or near-threatened, needed *both* natural forest with ample and varied dead wood nearby *and* large areas of old forest in the surrounding landscape in order to emerge.

This is actually logical and well known. The whole point of managed forests is, precisely, to remove wood for use as building timber, paper, vanillin and other goods, as described in previous chapters – and of course that means fewer trees are left to die in the forest. We fell the forests and remove almost all the trees from the logging site. That means less variation in dead wood and, in particular, fewer of the strange and rare types of dead wood – like fallen trees so large in diameter that it would be difficult for you to stride over them, or dense wood from stunted trees suppressed for a century in the understorey. The consequence is that fungi or insects that must have strange habitats like these in order to get by disappear from the forest landscape.

It is estimated that in a Scandinavian conifer forest landscape untouched by humans, some 60 to 80 per cent of the forest would be truly old-growth natural forest, over 150 years old. The rest would be natural forest in a variety of younger stages, the result of forest fires and other disturbances that are a natural part of our forest ecosystem – but in these parts too there would be many old trees that had managed okay, as well as large numbers of dead trees.

Nowadays, we fell the forest long before it gets that far: less than 2 per cent of Norwegian forest today is in a condition even resembling natural, old-growth forest. The vast majority of the forest, around two-thirds, has already been clear-cut once (or cut using similar forms of open felling). After 60–90 years – much less than halfway through a spruce or pine tree's natural life – it is clear-cut again. Clear-cutting is also eating its way into the remaining third, which was selectively felled in the 1800s, but has since been left in peace because it is inaccessible, at a great distance from roads.

Most people hiking through the forest don't know this. It is yet another example of shifting baseline – when only a thousandth of Norwegian forest is true old-growth forest, it goes without saying that few Norwegians alive today have wandered through such a forest. We no longer know the forest as it looked before widespread, serious felling began; we think today's forests are normal. And it may sound great if I tell you that the number of dead trees has tripled since 1920. But if I add that *even so*, today's dead wood constitutes just a fifth of the level seen in an authentic old-growth forest – well, your impression will change at once.

I believe this drastic divergence from the natural dynamic of the forest is woefully under-communicated. We know far too little about the consequences of this – over the long term and on a large scale. What is the implication of this continuous impoverishment in the species diversity of these decomposers? If we as a society are to adopt a position on whether this deterioration is a price we are willing to pay, we must at least be honest enough to communicate what that price is. We must also tell people that there are good alternatives to the current logging regime: better compromises between the removal of resources and species diversity; forest management and logging methods that will increase our chances of getting this part of the caretaking company onside in future, as we face the prospect of hotter weather, higher rainfall and other changes; methods that will allow trees to continue falling in the forest, to be spun about with mycorrhiza, to become home to thousands of baby beetles, to be broken down and slowly swallowed up by the soil. So that springtails, mycorrhizal fungi, mites and bacteria can continue working down there, until the nutrients are absorbed again, providing nourishment for new shoots, new trees.

Decomposition and soil formation have gone about their business since life first established itself on land several hundred million years ago, and this recycling of nutrients is a prerequisite of life as we know it. My wish is for more people to grasp the significance of dead trees and their inhabitants, to see the beauty in a rotten branch. In the meantime, we scientists will have to keep going about our business, studying connections, documenting what happens in this brown food web, in the cycle that binds together death and life – fitting the pieces together bit by bit in one tiny corner right at the bottom of the vast jigsaw puzzle that is life itself.

Of Reindeer and Ravens

Ghastly grim and ancient Raven wandering
from the Nightly shore –
Tell me what thy lordly name is on the
Night's Plutonian shore!
Quoth the Raven 'Nevermore'.

EDGAR ALLAN POE

'The Raven'

It's like a surreal battle scene from *Game of Thrones*, high up on the plateau. Hundreds of headless corpses. Many piled one upon another in bundles, others scattered. But they aren't warriors. They're reindeer – 250 adults and 70 calves – slain in a flash. Because nature is not cute. Catastrophes are part of the natural order, of nature's dynamic, surging balance – and things can get rough when Mother Nature swings her sword.

As when a severe storm headed over the eastern parts of the Hardanger Plateau one Friday in August 2016. The crashes of thunder tearing the air above the high mountain plateau caused the reindeer flock already gathered on a slope between Mogen and Stordalsbu to huddle even tighter together, afraid now. Suddenly lightning struck.

Maybe fore and hind legs acted as poles, conducting the current through the animal. Perhaps its soaked hide was especially conductive. Whatever the case, that lightning strike killed 323 reindeer, perhaps the entire flock, in the blink of an eye. Now they lie here, abandoned to the ministrations of nature's cleaning crew.

In record time, my biologist colleague Sam – who usually does research on large predators – drums up a team of young scientists, gets the necessary permits, packs tents and equipment and heads off to study the mass death. They call the project REINCAR – a clever play on the words reindeer, carcass and reincarnation. The authorities have already attended the site and removed the heads of the animals to test them for infection (with chronic wasting disease, a feared cervid sickness). But other than that, the animals will be left there to let nature take its course.

The scientists establish monitoring plots, and mark up fixed squares of half a metre by half a metre, which are to be monitored. Some are in the middle of the battlefield, others slightly further off, so that it will be possible to see the differences between the ordinary mountain ecosystem and this island of carcasses. Microorganisms, plants, insects, birds and mammals will all be studied.

Although we humans may think carcasses and decay are disgusting, the decomposition of dead animals is a vital and necessary process in nature. A carcass is like a pop-up restaurant for carrion eaters and decomposers, an island offering an abundance of accessible nutrients for a limited time only. But competition is tough, so speed is of the essence: American studies from the 1960s found more than 500 species of insect and other bugs alone in piglet carcasses placed out in a South Carolina pine forest. In just six days, 90 per cent of the pigs were gone – although the speed of the cleaning work is, of course, dependent on both temperature and a number of other factors.

Things don't move as fast as this on the Hardanger Plateau, 1,200 metres above sea level, but the process is the same. Carrion-eaters of all sizes long ago flocked around the carcasses. Insects like blowflies and flesh flies. Sexton beetles – black and red beetles in the cleaning sector – but also mammals and birds, such as red foxes, Arctic foxes, wolverines, golden eagles, ravens. Several of these larger animals are omnivorous opportunists that can eat dead animals, insects and plant material. Carcasses are a life-giving resource for all of them. At the same time, all these decomposers play a crucial role in bringing the nutrients from the dead animals back into the circle of life.

But the effect of a carcass doesn't end once the body has gone. The ripple effects can extend over years, possibly decades. That is what Sam and the other scientists want to investigate further: what actually happens to the whole ecosystem, on a large scale and over the long term, when a lightning strike stops the hearts of an entire flock of wet reindeer on a mountain plateau?

Specifically, lots of ravens turn up, as if out of nowhere. The scientists see them afterwards, on the wildlife cameras that have been set up; there may be hundreds of ravens in a single image. This '*gaunt, and ominous bird of yore*', as Poe describes it in 'The Raven', comes from a long way off. Not from the '*Plutonian shore*', as in the poem, but from remote parts of the plateau. To eat. In their bellies, the ravens bring the remains of earlier meals and many of them have been gobbling down crowberries (which are, botanically speaking, stone fruits, with hard shells around their seeds).

Once digestion has run its course, the hard-to-digest seeds emerge in the bird droppings. Scientists have found viable crowberry seeds in nine out of ten raven droppings. The raven droppings are concentrated around the reindeer, where the crowberry seeds are presented with perfect conditions, as the mineral soil is now exposed beneath the dead animals – bare, black earth, with no competition from other plants. This is because the nutrient shock from the rotting reindeer has altered the pH and nitrogen content so rapidly that heather, dwarf birch and other plant life have vanished.

The first scientific article published about the reindeer carcass project deals with precisely this: the fact that such islands of carcasses can be important termini for seeds spread via animals. Because the ravens come flying in with seed from other places, this may also influence the plants' genetic diversity in the landscape – in other words, the way the genetic material of different plant populations is mingled over time.

But all this takes time. Crowberry grows slowly. The scientists count and measure the small crowberry plants that send up shoots, but expect to continue for many more years to come, also with other partial studies and MA dissertations. (Among the more curious 'results' from the study site is the discovery of a peso banknote – which fell out of the pocket of one of the Mexican tourists who were so curious that they simply *had* to take the trip to see the reindeer cemetery.)

Nowadays, several years after the lightning strike, nothing remains but hard-to-decompose bones. The ravens have gone. Like pale witnesses, the bone fragments show us how effective nature's master cleaners are. On top of the bones and crowberries stretches wavy hair grass – an ordinary species of grass that produces large quantities of reddish flower heads for a time in disturbed, nutrient-rich landscapes. The grass colours the study area with its characteristic red hue as if to mark a stage on the journey back to normal mountain vegetation. If you zoom in on Vesle Saure on a satellite image of the Hardanger Plateau, you can see traces of the recycling of 323 dead reindeer in the form of a small, pink patch. It's a definite case of 'nevermore' for a flock of reindeer that has now returned to dust.

CHAPTER 7

The Warp in the Tapestry of Life

Ready for a tongue-twister? Give this a try: *Prochlorococcus* – a name that means 'primitive green berry' in English. Behind this name that skips across your tongue hides a minuscule organism – a blue-green bacteria. But don't be taken in by the word *primitive*: we're talking about a tremendously significant, oxygen-producing life machine here. This green berry alone is responsible for an estimated 5 per cent of all the photosynthesis in the world.

If you stick your hand in the sea almost anywhere between the Arctic and Antarctic oceans and draw up a handful of saltwater, I promise you you'll be looking at *Prochlorococcus* in its hundreds and thousands. However, sized at around half a micrometre (smaller than a grain of pollen or a drop of hairspray), it is hidden from our gaze. And this microorganism was concealed from us for a long time, a surprisingly long time for such a crucial organism;

it was only discovered in the 1980s, by a scientist called Penny Chisholm, who dedicated her entire career to this green mote. Thanks to her research group, we now know that *Prochlorococcus* is found everywhere in low-nutrient waters, from the surface water all the way down to depths of 200 metres, where there is barely any light. It may well be the world's most populous organism, with around 3 billion billion billion individuals (roughly as many atoms as there are in a tonne of gold), every one of them busy absorbing carbon dioxide and producing oxygen absolutely all the time. Over half the oxygen we breathe is produced in the ocean.

Photosynthesis is fundamental to all life, and *Prochlorococcus* is a representative of nature's fundamental support services – the life processes underlying all the other goods and services, and necessary for ensuring a functioning ecosystem. These are maintenance services that are crucial for life on Earth itself: photosynthesis and primary production, the creation of habitats, the nutrient and water cycles, soil production, the regulation of harmful organisms. We might therefore describe these natural goods and services as the warp in the tapestry of life – the threads you stretch lengthwise across a loom when weaving. Millions of species and their habitats are woven in between the warp threads, creating nature with all the other goods and services we benefit from so greatly.

The difference between the services of the caretaking company and these support services is a question of time and degree – the support services act over longer periods and on a greater geographical scale. But the transition from one to the other may be fluid: while decomposition into soil occurs locally over dozens of days or years, we can talk about the build-up of fertile soil as a slow, global process that takes place over thousands or even hundreds of thousands of years.

Whale Fall and White Gold

Let's go back to the sea again: picture yourself floating in saltwater, pitch darkness all around. It is ice-cold. The temperature is just a few degrees above zero. Somewhere beneath you lie desolate stretches of ocean bed. Above, the upper layers of water, full of sunlight and life and movement, where *Prochlorococcus* and other plankton produce oxygen, serving as the cornerstone in a rich food chain. All those tonnes of water above you create immense pressure – as if you had a stack of fully-grown elephants standing on your head. Welcome to the deep sea, the inhospitable habitat 200 metres below the surface that covers a full two-thirds of the planet and contains 90 per cent of all marine life.

Sometimes, it snows down here. Snow that never melts. Marine snow. Not ice crystals but tiny flakes of dead organisms from the layers of water above, which provide much-needed food for life in the deeps. Once in a blue moon, something truly titanic descends from above: a whale fall. The very words make my brain tingle. In my mind's eye, I picture a vast mountain of flesh and blubber and bone sinking slowly, majestically, through the water masses. Tonnes of carbon, nitrogen, calcium, phosphorus taking life's last dive. I don't know how long it takes for a dead blue whale to sink to its final resting place, but between the moment it lands and the point when every trace has vanished, we're talking decades.

Food is in limited supply on the ocean bed at depths of a thousand metres and the gaps in time and space between potential meals are large. A whale fall in the deep sea is like a lavish hotel buffet in the middle of a desolate desert – a rich food source that the inhabitants down below are more than capable of exploiting. The whale becomes a hotspot for a strange, and to some extent unknown, diversity of species.

Let's take a quick look at one of the peculiar types you may find on the whale carcass. Bone worms, also known as zombie worms, are segmented worms, making them distant relatives of species like earthworms and leeches. But the family likeness isn't exactly striking: bone worms look more like plants, with a root-like structure at one end and brightly coloured, swaying feathers at the other. They eat bones but

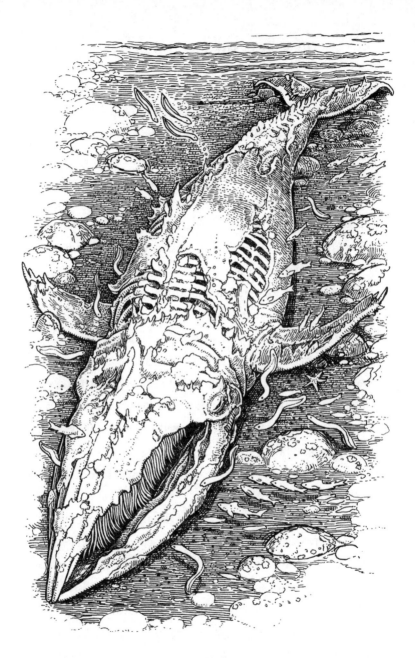

A dead whale in the deep sea is called a whale fall. The carcass is like a lavish hotel buffet in the middle of a desert – it supplies sustenance to deep-sea organisms for decades.

have no teeth – or even mouths. Instead, 'the root' secretes an acid that dissolves bone structures and releases the nutrients in close collaboration with bacteria that live in their bodies. The 'feathers' at the opposite end act as gills. Since this genus was discovered, as recently as 2002, there have proved to be many more species spread across all the oceans of the world. This has prompted some scientists to speculate whether bone worms also eat other things apart from whalebone.

Bone worms have a somewhat unusual sex life. The females are bigger than the males – so much so that in human terms, it would be like me having a husband who fitted in a teaspoon! And because it is desolate and lonely down there in the deep sea and difficult for them to find each other, bone worms keep things simple: the dwarf males simply live inside the female. Not just one of them, either, but a whole harem.

While we're on the subject of size and sex differences, by the way, the female blue whale is also bigger than the male. And since the blue whale is the biggest animal that has ever lived, this means that the planet's biggest individual animal ever must have been a female – a great mountain of a blue she-whale.

Although a dead whale is a good thing for the food cycle in the sea, a living whale is even better. Just as the trees of the Amazon are a biological pump that influences the never-ending water cycle, these vast whales fuel a pump system that moves food through the sea, both horizontally and vertically – by eating in one place and jettisoning waste matter or dying in another. In this way, they bring food to places where it is needed, creating vast ripple effects for other life.

Scientists have seen how the great whales, like humpbacks, sperm whales and blue whales, help pump food into those parts of the sea where it is most urgently needed. These whales dive deep into the ocean in search of different kinds of food – fish, octopuses or krill. They then swim up to the surface to breathe, and here they also expel their faeces, which float. In this way, the whale transports nutrients and minerals (such as nitrogen or iron) up to the surface water. In some ocean areas, such as the Antarctic Ocean, phytoplankton growth is restricted by access to iron. The iron concentration in sperm whale faeces is at least 10 million times higher than in the water so the whale's presence causes higher phytoplankton growth, which, in turn, means more photosynthesis, and the capture of more CO_2 from the atmosphere – carbon that tends to sink to the depths of the sea in the form of marine snow once the short plankton life has ended. A cautious estimate from the Antarctic Ocean suggests that sperm whales there send several hundred thousand tonnes of carbon out of the system and down to storage in the depths of the ocean every single year.

In addition, many of the great whales embark on long journeys, among the most impressive annual migrations we know of in any mammal. The humpback whale, for example, grazes in cold, nutrient-rich water at high latitudes, then moves to warmer, typically nutrient-poor ocean areas closer to the Equator to calve. The whales don't usually eat while in the calving areas, they simply live off their blubber. But they do have to pee and the urine they expel is rich in nitrogen – often a rare commodity in these waters (and that counts when you are big: one Icelandic scientist estimates that an average fin whale expels 974 litres of urine every 24 hours ...). So the whales' long migrations become part of a food conveyor belt, from rich seas to nutrient-poor ocean areas.

Nutrients are also moved further – up onto land and into freshwater, through salmon that travel upriver and die there, not to mention seabirds that catch their food at sea and poo on land (in fact, poo from penguin colonies is visible in satellite photos and is used to track them). It's hard to resist the temptation to describe this as a *shit-hot* food transport service because we're talking about enormous volumes here: every year, seabirds move 3.8 million tonnes of nitrogen and 0.6 million tonnes of phosphorus inland – a significant source of nutrition for land-based life.

Nesting seabird colonies along the coast are especially important. Here, the bird dung builds up over the years, acquiring a new and more beautiful name: guano. This word originated in Quechua, a language spoken by around 10 million people in the Andes of South America (incidentally, we have Quechua to thank for words like llama and cocaine too, and it also served as the starting point for the fictional *Star Wars* language, Huttese, as spoken by repulsive toad-like Mafia boss Jabba the Hutt). The Incas of South America used guano as fertiliser for hundreds of years before the Europeans arrived in the sixteenth century. Along the coast, each village had its allotted island where it could harvest guano, and severe sanctions were imposed on anybody who disturbed the birds while they were answering the call of nature.

Globe-trotting German naturalist Alexander von Humboldt (he who warned of falling Brazil nuts in Chapter 3, *see page 54*) was the person who brought the first samples back to Europe. He was extremely sceptical about what the locals had told him – that guano came from birds' backsides. There was simply too much guano for this to be true, he thought, and suggested instead that the guano mountains were the result of some mysterious catastrophe in the dim and distant past.

Europe's chemical engineers rapidly established that guano was super-food for cropland, as it was full to the brim with nitrogen, phosphorus and potassium – all of which are important plant nutrients. Around the mid-1800s, a short but intensive period followed in which guano was harvested in massive quantities. Entire islands were literally razed to the ground in the race to acquire 'white gold'. The US even introduced a special law, the Guano Island Act of 1856, which established that if an American citizen found an island of guano that no other country had laid claim to, this island could be viewed as American and its discoverer could remove all the guano he wanted and sell it – to Americans, that is.

Just as abruptly as the whole thing started, the bottom dropped out of the guano market; the white mountains of bird dung were gone: mined, carried off, spread over agricultural land, and grown into ears of wheat and potato tubers in Europe and the US. Even the larger seabird populations of the day couldn't produce enough poo to keep pace with the harvesting. Luckily – for our food production – artificial fertiliser was invented not so long afterwards.

In prehistoric times, massive amounts of nutrients flowed along a supply line from the deep sea to the surface, from sea to land and from coastal to inland areas. Nowadays, this flow has been interrupted. Most of the giant grazers are extinct, as mentioned earlier. Although some whale populations are now on their way back up, they are still well below the levels seen before we started hunting them. Many fish stocks have collapsed and seabird populations are in free fall.

Although the conveyor belt is still running, it carries only a fraction of the food it did previously: the marine mammals' transport of phosphorus from the deep sea has fallen to a quarter of previous levels, while the transport from sea to land by seabirds and fish travelling upriver has fallen by a full 96 per cent, according to research. The distribution of food from nutrient-rich to less nutrient-rich areas on land is deteriorating and, as a result, some ecosystems are starved of nutrients. At the same time, it is difficult to say precisely what the consequences might be because, logically enough, we lack data about how fertile the soil was 10,000 years ago.

The first stage of nature's great food conveyor belt – the whales' transportation of food from the deep sea to the surface – is especially important: if phosphorus and other nutrients disappear into the sediment of the deep sea, they are, in practice, out of action from our time perspective. Add to this the fact that we are currently emptying the Earth's stores of easily accessible phosphorus and we have yet another argument for restoring the populations of whales and other large marine mammals to earlier levels – as essential contributors to nature's flow of nutrients from sea to land.

The World's Most Beautiful Carbon Store

If I say the word 'carbon', what does it bring to mind? Barbecue charcoal, diamonds, the climate debate? Carbon is all this and a great deal more. It was created in the stars and is crucial to life as we know it. Look at yourself in a full-length mirror and what you see is roughly 14 kilograms of neatly packaged carbon. Breathe out and you contribute to the carbon cycle, because the carbon on our planet is engaged in an eternal round, from sea to land to air.

Storerooms are usually dull places. We have a lot of them in the depths of the building at NMBU where I have my office: shelf after shelf bathed in harsh light, storerooms whose concrete walls cast sharp echoes. But nature's stores are a different matter entirely. I've visited what I think must be one of the world's most beautiful carbon stores – the coast redwood forests of California. Seldom have I felt at once so small and yet so big.

Small because the tree trunks around you are so enormous, so utterly disproportionate that you can barely believe your eyes – as if you were a tiny mouse and the trees were elephants' legs. A wispy haze steals among the tree trunks and giant bracken like the mist we Norwegians call *alvedans* – 'elf dance'. If you tip your head back, you can just about spy a green bush of a canopy up in the heights, some 100 metres above you.

At the same time, there is something vast, spiritual about being here; about feeling a sense of belonging with the slow life around me. *The Word for World is Forest* is the title of a book by US fantasy writer Ursula K. Le Guin (1929–2018). And that is just what this feels like. I breathe slowly and hope that my exhalations will be captured by a conifer needle up there in the heavenly green bush; that 'my' carbon atoms will be moulded into bark and biomass and become part of the forest's world – part of the world's most beautiful carbon store.

Although coast redwoods are so enormous that it takes a small crowd of people with arms outstretched to reach all the way around their base, the tree trunks are not where you'll find most of the carbon. Well over half – indeed, possibly as much as 80 per cent of the carbon in the forests of the northern hemisphere – is found below ground.

Soil is simply a massive carbon store, regardless of what grows on top of it. Even so, it's a drop in the ocean compared to the sea. The sea is vast and stores much more carbon than soil, plants and atmosphere combined. Every single day, every single minute, carbon atoms are moved through this system with the aid of photosynthesis, combustion, decomposition, uptake in water and other processes in the never-ending carbon cycle. But nature has stowed away the vast bulk of carbon (well over 99.9 per cent of the planet's total, in fact) in a rather inaccessible store – buried in sediments, in the Earth's crust and in the Earth's core. That is what we are messing with when we extract fossil fuels such as oil and gas, releasing the carbon into the cycle above ground. The effects should be well known: although land and sea can absorb a lot of this, the carbon dioxide content of the atmosphere continues to rise too. Before the Industrial Revolution, the CO_2 concentration in the atmosphere was 0.0277 per cent. By 2017, it had climbed to 0.0450 per cent. That increases the greenhouse effect, making the planet warmer – with all the challenges that ensue.

The effects in the oceans are perhaps less well known. When there is more carbon dioxide in the air, more of it is absorbed by the sea. That decreases the sea's pH, making it more acidic. On average, the acidity of the surface water has risen 26 per cent globally since before 1750. The seas contain myriad species whose bodies are formed of calcium – from tiny plankton to enormous colonies of coral reef. Calcium is the white substance familiar to us from eggshells. When the sea becomes more acidic, its chemistry alters and calcium-based species struggle to produce their calcium shells. We do not know enough about the impact of the acidification of oceans but we do know that the Norwegian ocean areas in the north are especially vulnerable, in part because cold water can take up more carbon dioxide than warmer water.

As the saying goes, the devil is in the detail. That's a particularly appropriate expression to apply to the carbon cycle because, although the amount of carbon on the globe has remained unchanged since the planet was born and virtually all of it is still stowed away in the centre of the Earth, the whereabouts of the remaining gigatonnes is far from irrelevant. The extra carbon we are sending into the atmosphere via fossil fuels is precisely the kind of devilish detail that can have major consequences.

Healthy Nature Regulates Disease

Nature has inbuilt disease-regulating systems involving intricate interactions between different species. We would do well to familiarise ourselves with them. Increasing numbers of studies are showing that we can better ensure health – our own and that of both domestic and wild animals – by curbing the degradation of nature as well as securing intact ecosystems and the species diversity that goes with them.

Some have claimed that life boils down to a war against parasites. There are, at any rate, a lot of them, and infectious diseases are the cause of a quarter of all deaths on the planet. These are caused by many types of organism, with bacteria, fungi, viruses and various parasitic worms leading the pack. Most of these diseases, 60 per cent, can be transmitted between animals and humans – diseases like COVID-19, rabies, bird flu, Ebola (EVD), Zika fever, tick-borne diseases like Lyme disease and intestinal infections from bacteria, such as salmonella.

Over recent decades, ever more contagious infectious diseases have emerged, clearly dominated by animal-transmitted diseases, which account for a full 75 per cent of the new sicknesses. This is not a matter of chance. Our sweeping impact on the planet's ecosystems and climate is messing with nature's support systems for disease regulation and increasing the probability of transmission.

As an ever-growing human population destroys ever-increasing amounts of nature through agriculture, construction and fragmentation, we push ourselves into ever-closer proximity with wild animals. This increases the chances of encounters and contact between disease-transmitting animals and us humans, or between wild animals and our domestic animals. More than half of all new animal-transmitted infectious diseases since 1940 have been linked to agriculture and the food industry.

Legal and illegal trade in wild animals for food and medicine is, as I mentioned earlier (*see also page 96*), also implicated in the increased risk of transmission. Food markets where living and dead animals, wild and tame alike, are crammed together in wretched conditions are a challenge both to our health and animal welfare ethics.

Several of the serious infectious diseases transmitted by animals that the world has seen in recent years almost certainly resulted from the hunting and trade of wild animals. Not just COVID-19, but SARS, HIV, Ebola and variants of bird flu.

Our intrusions into nature also erode nature's inbuilt systems for reducing and regulating the risk of disease. These can take many forms: for example, some species are more suitable hosts for bacteria or viruses than others. The problem is that when we reduce species diversity, the species that are 'best' at transmitting infection are often the ones that flourish. Take small rodent species – generalists that thrive everywhere and adopt a kind of 'honk and drive' approach to existence: they live short lives and invest their energy in producing as many young as possible rather than a decent immune system.

Many large animals, such as predators, have very different life strategies. They live longer, relatively speaking, generally invest more in a solid immune system and can therefore often be worse hosts for many of the disease-transmitting organisms. In an intact ecosystem, these creatures act as a kind of buffer against the dissemination of disease because they 'thin out' the incidence of contagion. But large predators also need large areas to survive and do not thrive in close proximity to humans. That is why they are the first to vanish when we alter nature. And when they go, so too does their thinning-out effect. In this way, our impact can push conditions in nature towards both increased transmission and increased likelihood of transmission.

Nature's disease regulation is not just relevant to our health but also the health of plants and domestic animals. A study from Spain suggests that the presence of wolves can limit fatal animal disease such as tuberculosis among livestock. The study shows that when wolves take out wild boars, which act as a wild reservoir of the disease, this reduces the share of transmission without diminishing wild boar numbers. Instead of a wild boar population without wolves where many have tuberculosis and may die of it, there can be a similarly large wild boar population with less disease, whose population is controlled by wolves.

The advantage of a lower frequency of contagion is that the likelihood of farmers' domestic animals being infected becomes much lower too. One of the co-authors of the article points out that, although the wolves also kill domestic animals, the compensation paid to farmers is still only a quarter of the amount the authorities pay annually to combat animal tuberculosis.

Disease regulation is complicated. Although there are many questions outstanding, the IPBES is clear that conserving intact ecosystems and their native biological diversity will reduce the extent of infectious disease. Here is one important point: public health, animal health and the health of the environment are closely intertwined. Nature's systems inextricably link our health to that of our domestic animals, through the use of antibiotics for animals, through modern agriculture and vanishing species, through climate change. We have to think about this in a joined-up way, as the One Health concept emphasises. If not, we risk our children being the first generation to experience severe declines in both health and life expectancy.

Let's look at an example of how nature's disease-regulation can work. Once upon a time, passenger pigeons used to darken the sky for hours on end, so enormous were their flocks. The bird, then perhaps the world's most common, lived in North America and nested in enormous colonies in trees, living off acorns and other seeds. When you belong to a species whose individuals are numbered in billions, it goes without saying that you affect the entire ecosystem. But things changed fast. Scientists have estimated that the passenger pigeon apparently accounted for between 25 and 40 per cent (!) of all birds in North America before we humans took a toll on their population. In a matter of decades in the latter half of the 1800s, the species went from massively numerous to gone: extinct. The felling of the forests where the birds nested and ruthless hunting both played a role. With the invention of telegraph and railways, eager hunters could easily spread the word about where the colonies were nesting, travel there and send the birds they had trapped to market. Over the course of three months in 1878, half a million dead and 80,000 living passenger pigeons were sent by train from a single nesting ground in Michigan – and, apparently, the same number again were dispatched by boat.

That is sad enough in itself. But perhaps the extinction of the passenger pigeon also had other, unforeseen adverse effects. Because when billions of pigeons were no longer rooting around on the forest floor looking for food, the deer mouse – an American rodent that resembles the yellow-necked mouse – suddenly got a much bigger share of the seed buffet. Of course, we don't have annual data on the deer mouse population from the 1800s to date, but it is likely that this resulted in population growth. Deer mice fur is full of ticks and they act as a reservoir of tick-borne diseases, which are transmitted to us humans, like Lyme disease. One theory is that the extinction of the passenger pigeon is among the reasons why increasing numbers of Americans are falling ill with Lyme disease.

This is just a theory and can never be anything more. Since the passenger pigeon is gone forever, it is impossible to test the theory out, but we do have contemporary studies that support the link between nuts, deer mice and contagion. If there are a lot of acorns in a given year, there are also more deer mice the following summer – and with them come more infection-bearing ticks. Other studies show that where foxes and other predators keep the mouse population down, there are also fewer ticks. But this is a complicated interaction involving many more players and factors. The fragmentation of forests can play a role, as can the number of deer – and opossums too.

Most Americans aren't especially fond of opossums, despite the fact that they are the only North American representative of the marsupials – kangaroos and the like – have no fewer than 50 teeth and have the smallest brain in proportion to their weight of any mammal. What's more, people used to believe the male mated with the female via her nose (!) and she then sneezed her babies into the pouch on her belly – because the opossum has a forked penis, which appears to be perfectly adapted to Madam Opossum's nostrils. In fact, somewhat less visibly, the female opossum also possesses a matching bifurcated vagina and two uteruses. Anyway, despite this wealth of fun facts, most folk think opossums are vermin – ugly and rat-like.

Would it help if they knew that this half-pint pretty much hoovers up all the ticks in its vicinity? Because, as it turns out, the opossum makes extremely good use of the 50 sharp teeth it is born with – more than any other mammal. Out of six typical host animals for ticks, it was by far the best in class at picking off ticks and eating these stowaways. No fewer than 96 per cent of the ticks scientists placed on opossums in controlled studies were thus prevented from reaching humans.

We will never know whether the world would have been different – whether Lyme disease might have been less widespread, for example – if the passenger pigeon hadn't become extinct. The main point is rather to illustrate that usually we don't know the consequences of wiping out species, nor will we ever know what goods and services vanished with them. Because once they're gone, they're gone for good.

The Very Hungry Caterpillar

We scientists do a lot of peculiar things and creativity is a crucial component of good research. But if anybody had happened to pass the oak tree where Ross Wetherbee, one of the PhD students in our research group, was rigging up a new experiment in summer 2019, they might well have raised an eyebrow and wondered whether the fun had gone too far.

Ross's PhD is about all the insects that live in hollow old oak trees, and the contribution these insects make by providing natural goods and services in their surroundings. Some may live as larvae in dead branches or inside the oak's hollow, helping the dead wood become soil. As adult beetles, some of them fly from flower to flower, helping with pollination. Other residents of the oak may enrol in a kind of forest law enforcement agency; these are predatory insects that eat other insects and bugs, keeping them in check to prevent them from becoming too numerous. Because nature is wisely organised – with a dynamic balance between those that eat and those that are eaten. An eternal race where you have to hang on tight.

We would do well to understand these connections because many of the challenges we face with pests (and weeds) arise when we upset the relationship between prey (or plant-based food) and their natural enemies. As when we cultivate one plant over large areas, creating a banquet for caterpillars that enjoy this plant at the same time as we remove the habitat of the predatory insects that usually keep the caterpillar numbers down.

This was what we wanted to study. The idea was that the ancient oak trees acted as a source of and gathering place for these hungry predatory insects. In order to test this theory, Ross made artificial caterpillars. Fake caterpillars. Using green, brown and black Plasticine of the kind your kids play with in kindergarten, he moulded 720 x 3-centimetre-long pencil-thick caterpillars. Many were created in the café on the ferry between Moss and Horten, prompting funny looks and mutterings among his fellow passengers.

Ross left a tiny stump of steel thread poking out of each caterpillar so that they could be fastened to branches and twigs. Half of them were placed around an old hollow oak and the other half around a small young oak nearby, to ensure the forest would otherwise be as similar as possible.

When fake caterpillars are placed out in the forest like this, they are attacked by various predators that are fooled into believing that they are authentic caterpillar canapés. Since the caterpillars are fastened down, they can be collected again a few days later to check the bite marks, because the various insectivores – birds, mammals and insects – make totally different bite marks. The number of caterpillars with bite marks can be used to measure how many insectivores have been out and about.

These fake caterpillars have been used before. A major international study in 2017 set out almost 3,000 Plasticine caterpillars from Australia to Greenland to look for global patterns. And they found a systematic pattern: a caterpillar at the Equator is nearly eight times more likely to be eaten than a caterpillar at the Polar Circle, and it's not birds and mammals that account for this rise, but predatory insects – especially ants. That shows how important insects are as predators.

Ross's work is not yet complete but the interim results indicate that there are differences: more caterpillars were bitten around the old oak trees. A comparison of insect communities also confirms that there aren't just more predatory insects around the oldest trees but that they also display a greater range of traits. This makes the predatory insects' 'law enforcement agency' more effective and robust.

The moral of this example is a general point: in nature, it's all about eating or being eaten, and this happens in a kind of balance – a dynamic balance that is continuously evolving. The underlying support services in nature also include mechanisms to prevent individual species from dominating entirely – becoming mega-pests or super-weeds that take over the world. This is knowledge that we should be using to a much greater extent, especially in agriculture. If we collaborate more with nature in agricultural land-scapes, we can work towards win-win solutions, using nature's own systems to keep pests and weeds down – and at the same time using less poison and pesticides. This will allow us to cultivate at least as much food but in a more sustainable way.

Although most people are unaware of it, the support for this science is overwhelming and there are plentiful exam-ples. In more than 250 fields across the whole of the UK, for example, ground beetles helped out by eating up large amounts of weed seeds that would otherwise have become established in the fields, increasing the need for pesticides: more ground beetles meant fewer weeds. In Switzerland, farmers who sowed strips of flowers alongside their wheat fields found that damage from cereal leaf beetles, a major pest for wheat, fell 60 per cent – because the strips of land left to nature gave the cereal leaf beetle's natural enemies a place to live. In France, scientists compared almost 1,000 ordinary farmsteads of various types. They found that 94 per cent of the farms could produce just as much with far less use of poisons – and that two-fifths would, in fact, see increased production. In the tropics, retaining large trees above coffee and cacao bushes increases natural weed and pest control, as well as offering a number of other advantages – like increased profitability over a longer period of time. Across the UK, France, Germany and Spain, an experimental study is showing that variation and small units in the agricultural landscape boost the numbers of wild pollinating insects and increase seed production; we should therefore seek to limit the scope of large intensively farmed fields.

I could go on and on. If you also factor in that various weed or pest control substances have numerous undesirable effects that make the ecosystem less robust, resulting in the poisoning of nature, animals and humans (including several hundred thousand deaths every year, primarily in developing economies, according to the UN), the conclusion is self-evident.

The time is more than ripe to start using nature's services to secure good crops. We can do this by restoring more varied agricultural landscapes, in which natural vegetation, flower meadows and old trees – like the hollow oaks Ross is researching – shore up nature's own systems for preventing individual species from taking over, while simultaneously reducing our use of pesticides.

CHAPTER 8

Nature's Archives

Without libraries what have we?
We have no past and no future.
RAY BRADBURY

If you get off the Oslo Metro at the Sognsvann terminus
and glance east, you'll catch a glimpse of a white building
amid the pine trees. It's the National Archives, the four-
storey-deep, nuclear-bomb-proofed vault where Norway's
history is documented – from the painter Edvard Munch's
will to hand-coloured maps of the main thoroughfares
along the River Vorma from 1769, complete with neat notes
written in ink based on insect gall. Here, you'll find books,
documents and microfilms, along with more than 6 million
photographs and around 100,000 maps and drawings. In
sum, all this material gives us a picture of the changes that
have taken place and their underlying causes.

Nature also has archives. They take totally different forms, though: pollen deep in a bog can tell us when different trees and plants established themselves in Norway after the Ice Age. Ice core samples from the Greenland ice reveal how the climate has changed over tens of thousands of years. By comparing growth rings in standing dead trees and the timber of old buildings, scientists are able to painstakingly piece together a growth-ring archive that tells of growing conditions, logging and forest fires in earlier times. Corals, mussel shells and otoliths (or ear stones) in fish are other examples of materials where growth can be read off as different zones. In this chapter, we will look at some examples of nature's archives and what we can read from them.

When Pollen Speaks

To see a world in a grain of sand,
And a heaven in a wild flower,
Hold infinity in the palm of your hand,
And eternity in an hour.

WILLIAM BLAKE

'Auguries of Innocence'

Pollen is about more than plant sex and hay fever, it is also a source of information about the climate and vegetation of earlier times, about where we can find oil, about what Stone Age people ate. What's more, pollen can expose forged paintings and fake medicine, determine where honey comes from and help solve crimes.

Several hundred thousand plant species produce pollen; the grains are minuscule, and enlarged images of the different types remind me of kids' cereal – all chunks, spheres and ovals, their surfaces patterned with spikes, pores, warts, folds and carvings. Some pollen grains look like coffee beans, lemons or golf balls, while others are reminiscent of the now-notorious image of the coronavirus. Not all pollen is yellow, either, which is easy to see if you take a look at the hind legs of a fully laden bumblebee or honeybee. If, for example, it has gathered pollen from a Scilla in a spring-time garden, the pollen grains will be blue, while the pollen of common heather and raspberry comes in different shades of grey.

In the context of archives, pollen has a lot to offer: different plants have different pollen, so experts can recognise pollen grains by family or genus, sometimes even by species. The surface of the grain is made up of some of the most resistant substances known to us in nature, impervious to both fungi and bacteria. As a result, pollen preserves well, not just in bogs and at the bottom of seas and lakes, but also in fossils. And finally, plants produce large amounts of pollen that are usually released over a short period. It's not called pollen rain for nothing. Pollen grains can be spread by wind and water, attach themselves to animal pelts or the sole of a shoe – basically, they are almost everywhere.

That is how pollen, along with other tiny, durable particles like fungal spores, the remains of insect shells or soot particles from fires can help us create a picture of how the world looked before – and serve as a yardstick for conservation today. This area of specialisation is called palynology, which means 'the study of dust' in Greek.

To take one example: what did the old-growth, natural forest in Europe look like before we humans encroached upon it? Was it dense and dark beneath closed canopy cover, as the protected Białowieża Forest in Poland and Belarus is today? Or perhaps open, park-like woodland because large grazing animals kept small trees down – more like an English deer park or the oak-dominated landscapes near Linköping in Sweden?

To find out more about this, scientists use hollow bores to obtain core samples from bogs or lakebeds. The layers of the sample are like pages in a book where pollen grain and other 'dispersed dust particles' are the words. The latest studies argue that forests used to be more open in previous interglacial periods until we humans wiped out the large grazers, but that the canopy cover closed and the forest grew dense in their absence. This book is hard to read, though, and leaves much room for interpretation; the last word has hardly been said in the debate about Europe's old-growth natural forests.

Since pollen and other organic 'dust particles' tell us something about where a thing or person has been, they are also used in criminal cases, to uncover everything from forgeries and theft to assault and murder.

In New Zealand, a prostitute was brutally murdered in 2008. Months after the senseless crime, the police still didn't have any good leads despite an extensive investigation and hundreds of interviews. The finger of suspicion pointed to an organised gang, notorious for numerous crimes, which had a kind of clubhouse near the spot where the body was found. But the police had no proof linking the crime to this place – until they brought in a pollen expert. He discovered that the grass pollen found in the victim's nose had a special feature, an extra pore that could only be the result of a mutation – probably triggered by weed killer. The soft brome grass outside the clubhouse had been sprayed with weed killer. When the pollen from this grass turned out to share precisely the same unique appearance – and to be the only one out of many pollen samples from other potential crime scenes to do so – the police were able to establish that this was indeed the probable site of the murder. These details helped persuade a gang member to confess. He was eventually handed a life sentence for the killing.

A less gruesome example of the use of palynology involves a cargo of Scotch whisky that was dispatched to its destination by boat. To everyone's horror, the cases turned out on arrival to contain nothing but grey stone – limestone, to be precise. Somebody had removed the high-priced hooch – but where had it happened? Limestone was common both where the cargo came from and where it ended up, but closer analysis of microfossils in the rocks matched them with the bedrock close to the port the cargo was shipped from, establishing that *this* was where the police needed to search for the thirsty thief.

Rings of Lived Life

'So much life,' you think,
'So much secret life
these growth rings encircle!
A centre
Like the pupil in a seeing eye.'
HANS BØRLI
'From a Woodcutter's Journal'

The Norwegian poet and writer Hans Børli wrote melancholically about the way 100 years of patient growth must give way to the chainsaw's 'single minute of snarling steel'; about contemplating the stump and its growth rings afterwards – circle upon circle, traces of a secret life. Using modern methods, we can read these rings of lived life like the words in an archival document. The growth rings in living and dead trees tell us about climate change that contributed to the fall of Rome and reveal when grave robbers were on the prowl at the Oseberg ship burial mound. This kind of analysis has established the origin of a world-famous double bass and shown us that the altarpiece in Røst Church comes from a Baltic oak felled in the 1500s.

The bulk of a tree trunk consists of dead cells, cells that help keep the tree upright or contribute to transporting water between roots and canopy. The living part of a trunk lies between the wood and the outer bark. In this layer of growth, new cells form every growth season. Towards the outer edge, new cells are produced that will transport sap produced by photosynthesis. These cells eventually collapse, which is why this inner bark layer is thin in relation to the trunk itself. New wood cells are produced towards the inside and these are the ones that cause the tree's girth to increase.

Trees in temperate climes, here where we have distinct seasons, will grow most in spring and early summer, with growth abating in late summer and autumn. In this way, growth rings are formed – pale broad rings of spring wood; narrower, darker rings of autumn wood, ring upon ring. These rings are easiest to distinguish in conifers and certain deciduous trees such as oaks.

The breadth of the growth rings doesn't just change through the growth season but is also affected by variations in climate. Dry summers or cold, short summers yield lower growth and narrower growth rings. Trees that grow in the same area at the same time will therefore develop similar patterns, with alternation between narrower and broader growth rings, and we can use this to determine something about both when and where a given tree has grown. The study of growth rings is known as dendrochronology, which translates literally as 'the study of tree time'.

Like the National Archives near Sognsvann, every piece of wood – living or dead – contains a potential story for those able to read the language of tree time. Take the ransacking of Oseberghaugen, for example. Deep in this burial mound lay a powerful woman, entombed in 834 CE along with a Viking ship, exclusive grave gifts, 15 horses, four dogs and two axes – and what may have been her maidservant.

At some point or another between her interment in the mound and the Middle Ages, grave robbers came skulking around. They left six spades and four stretchers behind them in the earthen passageway into the burial chamber itself. Since these items were made of oak, it was possible to date the break-in by studying the growth rings in the tools. The only trouble was that the usual process is to saw a cross section of the wood, but such destruction of the artefacts was out of the question. Instead, they used a CT scanner like the ones used to take 3D images in hospitals – although this particular scanner was normally used to analyse rock samples. The analyses didn't yield the precise timing of the break-in, but did show that the stretchers were made from an oak that was still growing in 953. This, combined with the knowledge that the Vikings generally worked with oak wood soon after felling it – otherwise it became hard and difficult to fashion into tools – means that the date of the break-in can be narrowed down to sometime after 953 but apparently before 970.

The interpretation of age rings has many applications. Serge Koussevitzky was a world-famous Russian-born musician and conductor, best known for leading the Boston Symphony Orchestra over many years. He was the first person to translate Sergei Prokofiev's symphonic fairy tale, *Peter and the Wolf* (which has been narrated by figures as diverse as Eleanor Roosevelt, Sophia Loren and David Bowie).

One of Koussevitzky's favourite instruments was a double bass, which was said to have been made by the famous Amati brothers, Antonio and Girolamo, members of a family of phenomenal Italian luthiers. In 2004, the growth rings in the double bass were examined. It turned out that the spruce tree from which the top of the instrument was made was at least 317 years old, was still alive in 1761 and most probably grew close to the treeline in the Austrian Alps. Since the last of the two Amati brothers died in 1630, this ruled out the possibility that they could have made the prestigious instrument. It seems we must, instead, attribute the Russian's unique double bass to late eighteenth-century French instrument makers.

The dating of archaeological artefacts, buildings, musical instruments and art objects is exciting, but growth ring analysis also enables us to obtain valuable information about the interaction between nature and humans where historical sources fall short. Dramatic events such as forest fires, avalanches or rockslides also leave traces.

In Trillemarka, 80 kilometres northeast of Oslo – Norway's biggest forest reserve and my favourite hiking area – colleagues from the Norwegian Institute of Bioeconomy Research (NIBIO) have examined growth rings from almost 400 fire-scarred pine trees to understand the history of fires in the area. A fire scar is damage produced at the base of the trunk on a tree's leeward side when a blaze travels through the forest. Although the tree continues to grow, the damage is visible as a scar in the growth rings. The pine trees tell a story of both climate and humans: up until the early 1600s, forest fires were large, intense and mainly driven by the climate – by lightning strikes that set the forest ablaze in hot, dry summers. Over the next 200 years, however, the fires became numerous but smaller – a result of the growing population and the use of slash-and-burn cultivation. In the 200 years after that, up until almost the present day, the frequency of fires fell again, to the point where forest fires have now almost been eliminated – because slash-and-burn farming came to a halt over the 1800s as the forests became increasingly valuable for their timber.

So, we can read the trees' very own growth ring language to gain a better understanding of how the current composition of the forest was created, as well as which processes shaped the forest before we turned up with our chainsaws and felling machines. In this way, growth ring analysis can help us learn from history. By examining growth ring patterns from more than 9,000 European wooden artefacts and comparing them with written sources, scientists have shown how fluctuations in rainfall and temperature over the past 2,500 years have coincided with significant upheavals in pre-industrial society. During the golden age of the Roman Empire and the period of prosperity in the Middle Ages (roughly 1000 to 1200 CE), summers were warm with ample rainfall, whereas the fall of the Western Roman Empire and the troubled Migration Period coincided with an era of growing climate volatility, from around 250 to 600 CE.

Although today's society is better able to withstand variations in climate over the short term, our own society is not immune to fluctuations either. The secret language of growth rings tells how crucial climate is to a stable, prosperous society. Perhaps the tales told by such archives can give us even greater motivation to rein in human-caused climate change.

Chimney Talks Crap

Picture an old five-storey brick building in Canada with a chimney running all the way from the ground floor to the roof. Chimney swifts – a relative of the common swift that nests beneath your eaves – nest close to the top. Swifts are impressive birds: they are almost constantly on the wing and can eat, sleep and mate as they fly. But the result of that mating requires a nest, and in the case of this particular species, they are generally built in chimneys. They are simple constructions – a few twigs stuck together to form a kind of hammock-like structure, stuck to the wall with saliva. Here, the eggs are laid and here, the young spend their first weeks. Much eating and excreting is done. The chicks simply stick their backsides over the edge of the nest and go for it!

In this Canadian chimney of ours, that process went on for more than 50 years, from the time the chimney stopped being used for burning in around 1930 to the time when the top was closed and capped in the early 1990s. On the lowest storey, metres of bird droppings remained in the chimney flue, layer upon layer of them. It takes quite some creative free thinking to grasp that this is a treasure. A hidden treasure. Because the layers of swift droppings contain a time series that can tell us about the birds' diet over 50 years. And about the chemical pesticide content, such as DDT, in the food they have eaten.

At the very bottom of the chimney pipe was a little door that allowed the scientists to crawl in and set to work on a truly crappy job – digging their way up through a 220-centimetre-high wall of droppings. After two days' digging, they had shifted enough dung to be able to stand upright and take samples from the different layers. Some samples were used to identify the remains of the insects that had ended their days in a bird belly – fortunately, insects have very durable exoskeletons. In order to date the layers, the scientists measured levels of a radioactive isotope produced by nuclear explosions.

The samples showed a striking change in the birds' diet at the end of the 1940s, precisely the period when DDT started being used in Canada. The share of beetles plummeted at the same time as the share of an insect group called 'true bugs' – aphids, cicadas, *Heteroptera*, etc. – rose. Other studies have shown us that beetles are more nutritious but also more vulnerable to DDT than true bugs. Consequently, this sort of enforced change in diet can result in birds obtaining fewer calories per catch, making it difficult for them to gain enough nourishment. There may also have been fewer insects overall, but as in most other countries, few people have bothered to monitor insects in Canada. That makes it difficult to know for sure.

What we do know, however, is that chimney swift numbers have fallen. In Canada, the population nosedived 95 per cent between 1968, when people first started counting, and 2005; and the species is on the global red list as *vulnerable*, with a 67 per cent decline from 1970 to date. The bird poo archive in the chimney points to one possible explanation for the chimney swift's dramatic decline. Because although the share of beetles rose somewhat later, perhaps as a result of the ban on DDT, the share of nutritious insects in the birds' diet never returned to the level seen in the early 1940s. This is knowledge we would have been hard put to find any other way. That's how it is with nature's archives: although it consists of elements other than words written on paper, this archival material can still tell a tale. If we are able to read that secret writing, new insights await.

CHAPTER 9

An Ideas Bank for Every Occasion

I'm not trying to copy Nature,
I'm trying to find the principles she's using.
R. BUCKMINSTER FULLER, INNOVATIVE
AMERICAN ARCHITECT

My father was a fighter pilot and I grew up near military airbases in different parts of Norway. The unnatural sight of tonnes of steel lifting themselves off the ground with a roar fascinated me as a child and I remain amazed that it works to this day. I'm hardly alone in this. For thousands of years we humans stared longingly at the birds as they bore themselves aloft before managing to master the art of flying for ourselves.

Winged beings have been central to myths and religions across different cultures – the flying horse Pegasus, angels and dragons are a few familiar examples. We marvelled at the birds and tried to learn from them, but for hundreds of years, we were fixated on the idea that the wings had to flap up and down. As early as the end of the fifteenth century, artist and inventor Leonardo da Vinci sketched an 'orni-thopter' (from the Greek *ornithos*, 'bird', and *pteron*, meaning 'wing'), a kind of mechanical bird costume driven by muscle power, and several others tried their luck with similar contraptions over the centuries.

But flapping was futile: human bodies are too heavy, our muscles too weak. The solution came only after the German pioneer of aviation Otto Lilienthal (1848–96) grasped the principle of gliding flight in around 1890 after many lengthy observations of albatrosses gliding for hours with-out once beating their wings. On 17 December 1903, above a windswept sandbank in North Carolina, USA, we humans officially elevated ourselves into the world of the birds, bats and insects. While it is true that the very first successful flight in the Wright Brothers' flying machine lasted just 12 seconds and the distance travelled was shorter than the wingspan of a modern jumbo jet, this feat illustrated what we can achieve when we imitate nature's solutions, applying all our wit and wisdom.

More recently, birds have inspired train designers, and flies, data engineers. Other species point the way to smart materials or fewer traffic jams. There are also some surpris-ing and bizarre examples of how we have used creatures like dogs, pigeons and bats to do our bidding, in war and peace alike.

The world's millions of species harbour many as yet unexplored smart solutions. After all, nature has had billions of years to develop them. Some vital principles also distinguish nature's organic processes from our technical solutions: nature's organic materials are created at normal temperatures and pressures. Nature also uses the fewest possible resources and the least possible energy, recycling any waste – a genuinely circular system. By drawing inspiration from natural processes, materials and shapes, we can find smarter and more sustainable solutions to our challenges.

Sacred Lotus with Self-cleaning Surfaces

With clear melting dew
I'd try to wash away the dust
of this floating world.
MATSUO BASHO, HAIKU FROM THE 1600S

The rain came abruptly. Not a gentle, friendly shower but a persistent downpour. There are five of us in all: my husband, our three teens and me. We only have four umbrellas. Since this trip to the Kyoto Botanical Garden in Japan was my idea, it's only fair that I draw the short straw when it comes to umbrella distribution. I feel my summer dress starting to cling to me like a floral cotton wetsuit. At the same time, though, the rain gives us an unexpected bonus, such a fun phenomenon that it tests our mobile cameras' tolerance for rainfall to the limit.

We are standing by the Lotus Pond, a kind of shallow raised pool. Plants resembling water lilies stretch their soft green stems from the muddy bottom but these plants aren't content to lay their leaves neatly level with the surface of the water like ordinary water lilies. No, the lotus plant aims higher: like a kind of turbo variant, a water lily on speed, its stems grow up out of the water, lifting both leaves and pale pink flowers towards the rain, half a metre above the surface. This creates an aquatic version of a fairy-tale forest beyond anything to be found in Sherwood. I understand why the clever foldable map I got at the ticket booth (now being reduced to illegible cellulose mush by the rain) describes the pond as 'a forest of sorts worthy of an appearance in *Alice in Wonderland*.'

I don't see any sign of Alice but I do see something else that stretches the imagination. The raindrops that fall on the lotus leaves and blossoms simply bounce off. They dance around in the leaves like shimmering silver spheres before either running over the edge or gathering into a glittering pool in the centre of the leaf. Along the way, the drops carry off any dust and dirt, leaving the pale pink buds and the leaves shining clean, unsullied.

The self-cleaning leaves of the lotus plant have inspired the manufacturing of self-cleaning surfaces for products such as windows.

This is part of the reason why the lotus is viewed as a sacred flower in several eastern cultures. In Buddhism, the lotus represents purity of body, soul and speech, elevated as it is above the muddy swamp of desire. According to legend, Siddhartha Gautama, who won the honorific of Buddha and is famed as the founder of Buddhism, could walk from birth, and lotus flowers blossomed wherever he placed his feet. Buddha and some eastern divinities are often depicted sitting on a lotus throne. The plant has also earned respect for having the longest-lived seeds known to humanity: in 1982, botanists in the US managed to get a 1,288-year-old Chinese lotus seed to germinate. But what is it that makes the lotus self-cleansing? How can the dirt be washed away so effectively and can we copy this trick?

That is precisely what Wilhelm Barthlott, Professor of Systematics and Biodiversity at the Botanical Garden in Bonn, wondered. As early as the 1970s, he noted that the leaves of some plants always looked clean under the microscope. Were they especially smooth? The professor compared leaves from different plants using a scanning electron microscope, an instrument that provides both massive enlargement and extreme sharpness of depth. When he studied a lotus leaf under this microscope, the surface he saw wasn't smooth at all – quite the contrary, what he saw looked more like the inside of an egg box, with masses of tiny bumps. The surfaces of these bumps were themselves uneven.

And that is precisely what makes the lotus leaves self-cleaning: because of these bumps, the raindrops that land on the leaf barely come into contact with the waxen surface. Instead, the drop rests on the tip of the bumps, gaining extra support from cushions of air between them. It's a bit like a fakir lying on his bed of nails: because his body weight is distributed over more than 1,000 nails, none of them penetrate his skin. Since there is so little contact with the surface of the leaf, the drops easily roll off and since a speck of dust doesn't have much contact with the surface of the leaf either, it will easily attach itself to the drop of water and trickle off with it.

Barthlott spent many years developing the idea of self-cleaning surfaces and selling it to industry. Not until the 1990s was *Lotus Effect* registered as a trademark, patented and publicised in a scientific article. Nowadays, you can buy self-cleaning paint and self-cleaning windows. Today, scientists continue to research ways of making the structures more durable (a window normally has a longer lifespan than a lotus leaf) and to seek new areas of application.

Scientists are also delving deeper into nature's ideas bank in search of inspiration from other plants with 'rain-wear properties'. Lady's mantle, which gathers morning dew into a bead of water at the base of the leaf, is among them. Since the lady's mantle retains its drop long after the dew has vanished from other plants in the flower meadow, people in bygone days used to believe this water had magical properties. Heavenly drops from the lady's mantle plant were an essential ingredient for alchemists attempting to make gold – a fact still reflected in the plant's genus name, *Alchemilla* – 'the little alchemist'. The water could also heal sore eyes, people thought, but that involved getting the drops to run straight from the leaf into the eye, which isn't especially easy to do – and now we know why.

If you look at a greatly magnified image of a lady's mantle leaf you'll see that at the base of the leaf where the water gathers there is a forest of hairs, each with a lump at the end. The hairs are a bit like spiked maces – medieval weapons consisting of a spiked ball on a shaft. This structure holds the water slightly above the actual surface of the leaf. As a result, it doesn't heat up as easily when the sun shines on the leaf, allowing the drop to remain there for longer. Just why this makes sense for the plant we don't know, but these water-retaining hairs explain why it's difficult to tip the water off the leaf and get it to run where you want it to – into your sore eye, for example.

Many plants have various similar water-repellent or water-attractant surface structures – lupines, red clover and cushion spurge are among the other plants being researched. With better knowledge of these sorts of smart micro details from the plant kingdom, material technologists envisage being able to design solar panels or other surfaces that direct and control water as we wish them to, all by themselves.

Shinkansen – The Bird-beaked Bullet Train

It's easy to travel through Japan on public transport. Or it was once we understood that in the villages, you pay when you get off the bus, not on. Between the great cities races the ever-punctual high-speed Shinkansen train, through paddy fields and bamboo, so fast that it's hard to focus on the vegetation when it grows too close to the railway line. And it certainly is speedy – capable of almost 300 kilometres an hour. It took several attempts for me to succeed in filming a train passing the platform, because it was simply gone before I'd even had a chance to pull my mobile out of my pocket and switch on the video.

This very speed has caused problems. The early Shinkansen models had a blunt, rounded front. When they drove into tunnels, the air in front of the train became extremely compressed and when this compressed air was pushed out at the other end of the tunnel it made a tremendous boom – a bit like the sound of a fighter jet breaking the sound barrier. For those living along the train line, this was extremely unpleasant.

Fortunately, one of the engineers tasked with redesigning the train was a keen bird-watcher. He drew inspiration from the beak of the kingfisher. These beautiful blue-and-orange, bullfinch-sized birds – which sometimes turn up as exotic guests in Norway – dive for small fish and water insects in rivers and lakes. The kingfisher's powerful long beak, which narrows to a point, glides smoothly into the water almost without a splash. The engineers tested out different train designs and found that by copying the shape of the kingfisher's beak, they could reduce air resistance, power consumption *and* the tunnel boom.

The new rolling stock that is scheduled to come into operation from 2030 takes things a step further. Alpha-X, the new Shinkansen model, will have a top speed of close to 400 kilometres an hour. The kingfisher-style aerodynamic front of the first carriage will be longer, measuring a full 22 metres. In this way the designers hope that the lessons learned from nature will ensure that these even speedier trains will operate without the annoying noise.

Birds, more specifically owls, can also help us with design ideas for less noisy planes. Owls, with their sharp beaks and soundless flight through the night, are enveloped in myth and mystery. In European culture, from Æsop's Fables to the Winnie-the-Pooh stories, they symbolise wisdom, while several North American indigenous groups regarded them as messengers from the kingdom of the dead. That last part isn't so surprising, perhaps, since owls are so soundless that they seem to materialise from the very darkness.

But how do owls manage to fly so quietly? Apart from having a wingspan that is large in relation to their body, which enables them to reduce the number of wing beats, the answer lies in tiny but crucial details in the bird's feather structure. The wing feathers have comb-like teeth or points on their leading edge to break up turbulence that would otherwise create sound, and soft fringes on their trailing edge that muffles the sounds even further. The whole body is also covered with the softest, most sound-absorbent feathers, which are quite gorgeous to touch. I got to try this one late autumn a couple of years back, when I had an opportunity to accompany licensed owl ringers on their rounds. The experience of holding a boreal owl for a moment until it had been properly registered, then seeing it vanish back into the night on silent wings, was nothing short of magical.

Nowadays, noise-muffling technology inspired by owl feathers is used on the blades of electric fans, and researchers are currently working on upscaling it to wind turbines and planes. Not only can bird-inspired design reduce the noise of air traffic, it can also cut fuel consumption. Perhaps if I go back to Japan in a few years' time, I'll travel there in a feathered, electric aeroplane.

I can dream, can't I?

Colours That Never Fade

One July day in 2018, an email fluttered into my work account right in the middle of the holidays. Three photos of a huge shimmering blue butterfly were attached, with a ruler for scale. The email was from a woman in Østfold County, southeast Norway. She was wondering if I could tell her what species it was. The butterfly had flown into her mother-in-law's bedroom. She had tried to help it to freedom but later found it lying dead on the floor. In itself, this was hardly an exceptional event – but the mysterious thing was that a glance at the attached images revealed the butterfly to be a male *Morpho*, a genus that lives in South and Central America. We may have had a tropical summer in Norway that year, but that didn't explain why tropical butterflies were suddenly turning up on the wrong side of the ocean.

Morpho butterflies are truly a wondrous sight. The species in this genus are among our largest butterflies, with wingspans of up to 20 centimetres. But the really special thing about them is the fantastic metallic blue sheen on the dorsal side of many species' wings, a play of colour whose nuances shift depending on the angle you see it from. The underside of the wings is also beautiful: brown with big circles that look like eyes.

But the blue surface of the wings is not actually blue – it contains no blue pigment. The colour is created by tiny structures in the scales on the wing's surface. What we're talking about here is nanostructures, sized at roughly a millionth of a millimetre or thereabouts. If you could zoom in and see details at this scale, you would discover that each scale is covered in tiny ridges. In cross section, these ridges look like Christmas trees, with branches pointing out to the sides. These nanostructures break up the light, reflecting it in special ways so that the surface gives the appearance of a shimmering metallic blue. In addition, semi-transparent scales lie on top of these structural scales, helping to spread the colour.

There are many attractive aspects to the *Morpho* butterfly's way of looking blue. The colour is intense and shimmering, discolouration is not a problem and because no pigments are involved, the colours last forever – they don't fade. As a result, there has been no shortage of research seeking to imitate the celestial blue of the butterfly wings. The textile industry is interested because it wants to produce fabrics with exciting properties. Use of these colours will also help reduce the industry's problematic reliance on poisonous dyes. The printing and security sectors are also keen: such techniques will, for example, enable the colour-coding of banknotes, making them almost impossible to forge. The technique can be used to produce more efficient solar panels or extremely precise chemical sensors.

Some have already tried out butterfly-style colours in practice. Back in 2008, French company Lancôme produced a cosmetics range called L.U.C.I. – Luminous Colorless Color Intelligence Collection. This patented invention involved blending colourless particles with structural colour properties into make-up to create what the company itself described as 'a pure and intense halo of colour' – 'a spectacular change' no less. As far as I can see, the range is no longer available. Perhaps the price was on the spectacular side too.

There are also *Morpho* textiles, developed in Japan under the Morphotex brand. Here, the fibres consist of many dozens of nano-thin layers of nylon and polyester, put together in such a way that textiles can be produced in red, green, blue and purple without using a single drop of textile dye. But for now, the challenge is to find cheap and efficient ways to scale up production of structurally coloured materials. There is a steady flow of new patents but much of the development is happening behind closed doors because of competition.

In the Amazon rainforests, the *Morpho* butterflies flutter around as before, blissfully ignorant of the stir their ingenious, natural and smart solution has caused in the patent sector. For them, the colour apparently serves as a signal to enemies to keep their distance as they flutter confidently above the tree canopies, watching over their tiny territory. Many people know about the *Morpho* butterfly from Gert Nygårdshaug's novel, *Mengele Zoo*, whose protagonist – a young boy called Mino – is a butterfly collector living in the rainforest. One day, the paramilitary turn up in Mino's village. They have come to clear the way for the oil industry and will stop at nothing to achieve their aim.

In the real-life Amazon, *Morpho* butterflies are under threat from the destruction of habitats and illegal collecting. These creatures are regulars at butterfly houses all over the world, and plenty of people are keen to have a patch of the sky for decoration – one of these butterflies on a pin. Large-scale breeding operations seek to satisfy this demand.

And this, then, is the explanation for the mysterious *Morpho* in Østfold: it turned out that somebody at a neighbouring holiday cottage had a visitor from Costa Rica and their guest had brought an exotic gift: four tropical butterfly pupae. They had hatched into adult *Morphos* and one of them, a beautiful blue male, found its way out into the Norwegian summer and from there, into the neighbouring cottage – where, as it happened, the son of the house was halfway through reading *Mengele Zoo*. The Østfold *Morpho* concluded its brief life there in the mother-in-law's bedroom. It ended up neatly fixed on a pin, a souvenir of a long journey and a curious coincidence.

Moths with an Eye for Darkness

Red-eye isn't especially flattering in photos. That irritating red dot is caused by the reflection of the flash on blood vessels at the back of the eyeball; instead of producing a portrait full of Christmas cheer, the compact camera flash creates a horror movie vibe. If you were a moth, though, that kind of reflection from your eyes wouldn't just be ugly but downright dangerous too. If the feeble light of dusk is reflected in your eyes, it's like switching on a lighthouse that guides all your predators straight to you. That's why many moths have a special anti-reflective layer on the surface of their eyes. We can copy this to make better surfaces for mobile screens, camera lenses and solar panels.

We have known about the moth-eye effect since the 1960s but it hasn't been easy to replicate. Again, it's a question of nanostructures: tiny little bumps shorter than the wavelength of visible light that cover the surface of the eye. These nano-bumps smooth out the transition between air and eye so that the incoming light is not reflected but continues straight through the material. This means, for example, that you can take pictures of a window display without seeing your own reflection. Or that it's easier to read the screen of your mobile phone, your car's GPS or the big departure boards at airports.

Better knowledge about the production of nanostructures has resulted in at least two Asian companies producing this kind of film ready-made in a roll, so that you can fix it to any surface you wish to. If you cover a transparent glass or plastic surface with anti-reflective film, 100 per cent of the light will pass through, against 92 per cent without the film – according to the publicity. The material is also water-repellent, like lotus leaves, and the manufacturers boast that they have now managed to make the surface durable.

These kinds of commercial nano-films have also been tested underwater. It turns out that many animals in the sea, like octopuses or sharks, have teensy-weensy structures on the surface of their skin. We do not entirely know what role these nanostructures on marine animals play – perhaps they help prevent light from being reflected underwater too, or reduce water resistance when they swim. Another function could be to prevent other organisms from latching onto them – and it certainly is the case that small, sticky sea creatures, indeed even bacteria, are less able to attach themselves to these nano bumps than to ordinary smooth surfaces.

That is very good news. Not just in water, where smart surfaces like this can reduce growth below the waterline on boats without the use of unpleasant chemicals, but also in the human body. Currently, such bio-inspired surfaces are also being tested out on items like tooth and bone implants, as well as urinary catheters, to reduce the growth of bacteria.

Smart as Slime Mould

Our modern, high-tech society poses a myriad of complicated challenges with many possible solutions. Just think of the logistics branch; about the way it has to distribute packages in vans and choose which routes the vans should drive. This is where modern computers' capacity for calculation comes in handy, but knowledge about nature's ways of solving similar problems can also offer useful inspiration.

Take ants. There are no traffic jams in the ant world. Even when they cover 80 per cent of the available space, they potter around efficiently without colliding and without having to stop – a feat we humans cannot imitate in similar crowd densities. In a study of 35 ant nests containing hundreds of thousands of Argentine ants, scientists made small bridges in different widths for the ants to cross in both directions, and set up surveillance cameras to study the traffic in slow motion. The recordings showed that every ant constantly adjusts to the traffic around it – and the way they do so alters according to density. They speed up if it's a bit crowded but take the pace down again and stop greeting each other and taking U-turns when it gets really crowded. Totally without traffic lights or roundabouts, they achieve a traffic flow we can only envy. The hope is that some of this can be mimicked in new driverless cars, making us humans as street-smart as the ants.

There are, in fact, plenty of algorithms based on nature and species – birds flying in flocks, shoals of fish behaving like a single organism. My son, who is studying to become a data engineer, recently drew my attention to the fact that there is a fruit fly algorithm, modelled on the way these cute red-eyed critters search for food. Not to mention the honeybee algorithm, based on the bee's decision about whether to perform a waggle dance and get her bee-sisters to gather more food in the place she's just come from or to return there alone.

Nature has had millions of years to solve complex problems, and ideas and knowledge may await you where you least expect them. Even in the simplest organisms, like slime mould. Ever since childhood, I've nursed a kind of crush on slime mould. First, simply because it looked so lovely out there in the forest, with its intense colours and cool Norwegian names like troll butter and witch's milk (the first of these has a decidedly less magical name in English: dog vomit slime mould). Later, as a student, I was captivated by the story of a slime mould scientist who left the object of his study safely stowed in petri dishes overnight – but found to his horror on returning to the lab the next morning that his mould had escaped.

The thing is, although slime moulds are decomposers with external digestive systems, they aren't fungi. And although they can move, gather and separate, they're not animals either. Nor are they plants, although they may produce something resembling flowers.

Victims of bullying systematics, slime moulds don't get to hang out with fungi, plants or animals, but are banished to the kingdom containing algae and single-celled species, and forced to play with seaweed and amoebae. This may be unfair: while slime moulds may not have brains (though they *do* have several hundred sexes, or mating types to be more precise), they turn out to be capable of performing surprisingly advanced actions.

You can, for example, place a bright yellow, sticky slime mould of the *Physarum polycephalum* species in the middle of a maze with many blind alleys and set out a little slime mould snack, like oatmeal, at the end of some of these. The slime mould will send tiny threads into all the passageways in search of food. After a few hours, it has discovered the shortest route to the food dish, recalls all the other threads and what you will see in the maze is the most energy-efficient route.

In 2010, Japanese slime mould scientists used this to show that slime moulds can match the planning abilities of human engineers. They made a kind of miniature map of the Tokyo area, placing oatmeal on the sites of the largest cities in the region. By varying the lighting on the map, they replicated the placement of mountains and lakes and other physical obstacles in the way of transport arteries (slime mould avoids bright light). Next, the scientists positioned a dollop of slime mould in the location of the capital and waited. In less than 24 hours, the slime mould had completed the task, connecting the oatmeal towns in the most effective network – which bore an astonishing resemblance to the actual railway system.

Later, the slime moulds were presented with several similar tasks. In a scientific article snappily titled 'Are motorways rational from slime mould's point of view?' the dollop got to compete with engineers nationwide when it was tested on no fewer than 14 simulated oatmeal maps from all over the world. Belgium, Canada and China are the countries where the slime mould's answer came closest to the actual motorway network – but that raises the question of who actually built the optimal solution.

Slime mould, bees and ants are hardly going to steal my engineer son's job once he's finished his training. But perhaps we can still learn some maths tricks from ants, bees and slime mould, which we can then use to produce more efficient or energy-saving networks.

The Hermit Beetle and the Hound

When the Man woke up he said, 'What is Wild Dog doing here?' And the Woman said, 'His name is not Wild Dog anymore, but the First Friend because he will be our friend for always and always and always. Take him with you when you go hunting.'

RUDYARD KIPLING

'The Cat That Walked By Himself'

Nature's ideas bank also encompasses our interactions with our pets and the different ways we humans can use animals. I am, myself, a happy dog owner. Well, not so much dog owner as carer: I have a dog from a guide dog school living in my home so that it can get used to family situations and everyday life. Sometimes I'll care for a puppy waiting to grow old enough to be tested and trained. Other times, it'll be a dog already in training that needs a host family for the holidays.

Having a pet can make you both happier and healthier. A soft-furred, tail-wagging golden retriever or an affectionate pussycat can help reduce stress and improve mental health. Pets may also lead to new social interactions, or make us go for walks. But animals can make themselves useful in quite different ways too – in both war and peace.

After insects, I think dogs must be the coolest creatures alive. Clever, patient and almost always good-humoured. What's more, they have a totally fantastic sense of smell. Place a dog at a right angle to a trail where a human has recently walked. After sniffing fewer than five footprints, the dog can work out which direction the person went in.

Because of this, dogs are used to find injured game, smuggled drugs or diseases in humans, as I already knew. But it was news to me that they can also be used in conservation – until I read an article about beetle hounds: dogs that are used to find rare and threatened beetles in old hollow trees. Could it get any better than that? This story has pretty much everything I love: insects, old trees and dogs too.

It turns out that some Italians have trained up an 'osmodog' – a dog capable of sniffing its way to hollow trees inhabited by the globally threatened hermit beetle. Usually, hermit beetle larvae are found by searching through wood mould – that lovely soft blend of rotten wood and fungus found inside hollow trees. The larvae really dig gnawing their way through this wood mould and the slightly rotten walls inside an old, hollow tree.

The disadvantage of searching the old-fashioned way is that it takes time and you risk harming the larvae as you search. But an osmodog – a specially trained beetle-seeking hound – really speeds things up. The dog locates the rare beetle in less than a tenth of the time it takes to sift through wood mould: all it takes is a few sniffs around the tree. If there's a whiff of hermit beetle larvae in the air, the dog sits down nicely and barks.

To be fair, there's no real need to dash off and train your four-legged friend to find hermit beetles if you happen to be in Norway. Here, there's only one place you'll find the species: in the small city of Tønsberg, in southern Norway. We believed it was extinct but then it turned up in a churchyard and today, it is listed as a prioritised species under the Norwegian Nature Diversity Act.

But perhaps it's some consolation that even we humans, with our pitifully poor sense of smell, can pick up the aroma of adult hermit beetles. If you wander through the old churchyard in Tønsberg on a late summer's day and smell a faint scent of peaches, well, that means love is in the air, because adult hermit beetles use a scent substance called gamma-decalactone to find each other and mate. It smells like sweet fruit, with a hint of peach or apricot. We use the same substance in cosmetics and food.

If you still want to use your dog to contribute to conservation, there are always alternative options. More than

1,000 other insects in Norway are under threat and many of them probably have a species-specific scent that your dog can be trained to recognise and search for. Dogs can also be used to track down dung from rare species, to locate alien species or to find bats and birds that have been killed by wind turbines. In Chile, clever Border collies run around carrying specially constructed pouches to spread seeds and get vegetation to grow back more quickly in fire-damaged areas, and in Iowa, man's best friend sniffs out threatened pond turtles. If none of that appeals, why not just take a hike together in the forest, you and your pooch?

Like a Bat into Hell

One of my absolute favourite childhood books was *The Brothers Lionheart*, the children's fantasy novel by Astrid Lindgren – a magical blend of brotherly love, loyalty and courage in the face of lust for power, evil and dragons. Do you remember the carrier pigeons Jonathan sent between Cherry Valley and Wild Rose Valley? Well, pigeons – and other creatures – have also helped out in conflicts in the real world.

Take Gustav, for example, otherwise known as NPS.42.31066: a grizzled carrier pigeon that flew the first message about the Allied landings in Normandy safely back to base in England. In return for this heroic feat, he later received the Dickin Medal, the highest distinction an animal can win for military or civil defence services. The bronze medal, inscribed with the text WE ALSO SERVE in capital letters, has been awarded to a total of 32 carrier pigeons, 34 dogs, four horses and one cat. It was, incidentally, last awarded as recently as 2018, to Kuga – an Australian military dog who saved the life of an entire company during an ambush in Afghanistan in 2011.

During the Second World War, plans were also afoot to use pigeons to target bombs. An American behavioural ecologist proposed having a special cockpit for pigeons at the front of the missile. The birds would be trained to peck on a screen displaying the bomb target, and cables attached to the pigeon's head would steer the bomb to its target. While the pigeon project never got off the ground, another war project involving the use of animals did – in this case, bats.

A couple of years ago, I was in Hiroshima. Apart from the ruins of the Chamber of Commerce, which still stand as a memorial close to 'ground zero', the city centre looks like that of any other Japanese town. There amid the high rises and the trees in the park, it is difficult to imagine the suffering people here experienced some 70 years ago. Men in suit trousers and white shirts hurry purposefully to or from the office. Locals air their little lapdogs at the edge of the park and tourists brandish their selfie-sticks. Yet there's a heaviness in the air. Something is different: lower-key, more inward-looking. As if everybody is struggling to comprehend.

At the end of the park lies the Peace Museum. Here, you'll find the hard facts – more than I can bear to take in. And objects, like silent storytellers. The tricycle that belonged to Shin, who was playing outside his family home when the bomb fell. When his father found him beneath the ruins, he was still holding onto the red plastic handlebar of the tricycle. I see teacups melted by the heat and think of the Tarjei Vesaas poem, 'Rain in Hiroshima': '*As she lifted her hand/to reach for the teapot/there was a blinding light –/no more/everything was gone/they were gone ...*'

I believe everyone who visits Hiroshima asks themselves: What would be different if the US hadn't dropped the atom bomb? But few know that there actually was another plan, a plan that sounds quite insane but was tested and seriously considered as a viable alternative. A plan that would spread chaos but result in fewer civilian losses. A plan that involved thousands of bats – and a dentist with unshakeable faith in his own ideas.

Lytle S. Adams was a dentist from Pennsylvania. In December 1941 he went on holiday to New Mexico, where he visited the Carlsbad Caverns – home to an enormous colony of Mexican free-tailed bats, which present an impressive spectacle when they leave the cave in their millions at dusk.

A few hours later, the dentist heard the news that the Japanese had attacked Pearl Harbor and made a remarkable mental leap. What if thousands of these bats were equipped with tiny, self-igniting incendiary devices and dropped over Japan?

Astonishingly enough, the dentist's crazy idea was adopted as a military research project. It probably helped that he had good connections with Eleanor Roosevelt, the First Lady. Less than a week after Franklin D. Roosevelt had received a project outline from Adams, he sent it on through military channels, accompanied by a note which read: 'This man is not a nut. It sounds like a perfectly wild idea but is worth looking into.'

Two million dollars and several years were spent developing the 'technology'. Six thousand bats had to pay with their lives but the dentist wasn't especially bothered about animal welfare. He was well-nigh convinced that God had created these bats for precisely this project: 'The lowest form of animal life is the bat, associated in history with the underworld and regions of darkness and evil. Until now reasons for its creation have remained unexplained. As I vision it the millions of bats that have for ages inhabited our belfries, tunnels and caverns were placed there by God to await this hour to play their part in the scheme of free human existence, and to frustrate any attempt of those who dare desecrate our way of life.'

The US military's research resulted in the following recipe for a bat bomb: take 1,000 bats and cool them down to send them into hibernation. Attach a miniature incendiary bomb made of napalm and a delayed-action igniting mechanism to the loose skin on their chest. Next, place the sleeping bats in cardboard trays, which are in turn stacked in a one-and-a-half-metre-long metal box shaped something like a conventional bomb. The metal box is dropped from a plane and opens beneath a parachute. This gives the bats time to wake up. When they fly off, a 15-hour countdown starts before the napalm ignites. The idea was that, in the interim, thousands of bats would fly down and settle in nooks and crannies beneath the straw and bamboo roofs of Japanese houses.

Proof that the mechanism worked came when some fully armed bats escaped, setting fire to a hangar at the airbase where the tests were taking place. Nonetheless, the bat bomb project was never put into effect. The launch of large-scale production of a million incendiary bomb bats was scheduled for May 1944, but shelved just months before. Instead, the US military chose to focus all its efforts on completion of another weapon – the atom bomb.

It's closing time at the Peace Museum in Hiroshima. I'm the last one out of the room where you can watch and listen to video recordings of eyewitnesses. It feels strange, wrong in a way, to go out into the dusk and suddenly find yourself faced with the neon lights and the bustling traffic of the indifferent city. A bit like a betrayal of all those who died, all those who suffered, to shrug them off so easily.

It's hard to say whether Lytle S. Adams's hare-brained scheme to end the Second World War using bats would have worked, had it been put into practice. The dentist himself contended to his dying day that the blazing bats would have terrified Japan into surrender – without the enormous civilian losses caused by the atom bomb.

CHAPTER 10

Nature's Cathedral – Where Great Thoughts Take Shape

Early of ages

When nothing was

There was neither sand nor sea

Nor cold waves.

The earth was not found

Nor the sky above

Ginnungagap was there,

But grass nowhere.

FROM *VOLUSPÅ*, THE MYTHOLOGICAL

NORSE POEM, WRITTEN DOWN IN ABOUT

1200 BCE

In autumn 2019, I was in the US to talk about books and nature. On my bedside table, not just in the hotel in New York City but also in the small towns I visited further north during a free weekend tacked onto the end of my stay, I would often find a nature sounds machine. It's a kind of clock-radio – only here, in addition to radio stations, I also had the choice of various sounds from nature, like 'babbling mountain brook' or 'birdsong in the forest'.

I realise mountain brooks are hard to come by in the Big Apple but there's still a paradox here that astonishes me. What is it about nature that moves us so much we want to fall asleep and wake up to canned nature sounds, even as we barely bother to look after what nature is still left out there – the real thing?

We know that nature brings joie de vivre and life quality, offers inspiration and a sense of belonging. And this applies to the full spectrum of nature – from urban green spaces two blocks away to wide-open mountain plateaux miles away from the nearest road. It's about nature as a playground for kids, as a training ground or an outdoor space for reflection in a hectic everyday life. About just knowing that there's a wild forest out there somewhere, where regeneration and ageing are controlled by the laws of nature and not the hydraulics of the harvesting machine.

Spending time in nature can give us a sense of being part of something bigger than ourselves; part of everything that has emerged from nothingness. Some link this to religious faith, while others experience it more as a profound reverence for life itself – a fascination and respect for the intricate details of nature's vast interactions. It is this sense I am thinking of when I refer to nature as our cathedral.

When Notre-Dame burned in Paris in April 2019 a whole world wept, for this magnificent cathedral is both a thing of beauty in the present and an awe-inspiring legacy of times gone by. Many have personal memories of the place; my own is from an international youth summer camp in Kehl, Germany, which involved a weekend trip to Paris. We sneaked out into the gentle night rain without permission and sat on the wall beside the canal, singing and gazing up at rose windows and reliefs.

Just as we try to conserve our cultural heritage and are now repairing Notre-Dame after the fire, so too should we conserve our natural heritage and restore nature where we have degraded it. Because nature is not just useful. It also has crucial immaterial values, which are impossible to measure or set a price on.

This is also a question of ethics, of the intrinsic value of species: we have a responsibility to take care of nature. Because other species also have an independent right to fulfil their life potential: beetles, butter cap mushrooms and beavers. Big and small, ugly and beautiful, useful or not.

My Life in the Forest and the Forest in my Life – Nature and Identity

Sometimes out in nature, I experience an intense joy. A powerful, focused euphoria that sits deep in my chest and radiates out to every fibre of my being, making me want to cheer and weep. This feeling always comes to me in the forest. Not in a well-managed older industrial forest where the trees all stand in rows, equally tall and equally broad. That kind of forest may be pretty in its way, just as a wheat field in the autumn sun can be beautiful. But my wild unruly joy lies hidden in a different kind of forest, among downed logs and dead pines worn silver, beneath ancient spruces raising their ragged canopies to the heavens.

I know enough about forests to know that this isn't true old-growth forest. We humans have left our fingerprints everywhere in Norwegian nature and you can see them here too, if you know what to look for. And so it will be in the future too. The majority of Norwegian forests will be harvested. Even so, these pockets of protected natural forest mean a lot to me – to the person I know myself to be, to my identity. Many people have this: a place in nature that owns a little piece of their soul.

For me, the joy of being in natural forest isn't merely intellectual but sensual too. Forests are the play of light and shadow along a spectrum of green. Soft moss beneath my boots and textures, from the rough bark of a pine to the smooth trunk of a beech; colour and scent and sound. And an old natural forest has a special sound to it – a chord of life and death, a note that has sung through the forests for millions of years since the Carboniferous age. It is the sound of dead trees leaning against the living, so that their trunks and branches rub and rasp against each other, creaking and squeaking to the rhythm of the wind that rustles in the twisted canopies.

A fellow scientist, originally from Montana, once told me that forests in Scandinavia felt all wrong to him for the first few years after he moved here. He couldn't put a finger on what it was, but something was missing. One day it struck him: the missing element was these sounds! Because compared with the wild forests he was used to from America's protected old-growth forests, Norwegian industrial forests are silent. In forest plantations, there is no room for dead trees. All you hear is the hushed whisper of timber prices and future input volumes.

Forests are also scent. A clear-cut has a sharp tang of resin and crushed conifer needles, of black forest soil slashed open by the tyre tracks of a forwarder. In mature, natural forests, the aroma is rounded and soft and changes constantly as you walk. Above a bottom note of transitoriness and hope – of life becoming death becoming life – raw notes of dark green moss alternate with spicy hints of sun-warmed bark. Perhaps you'll catch a waft of coconut from a dead spruce, its bark still intact – and if you follow that scent you'll find that the source is a nubbly pale-yellow coating: a rare fungus called *Cystostereum murrayi*.

Once upon a time, all the world was forest. It was big and dark and dangerous, with wild animals and scary things that filled us with fear. We humans were only able to let down our guard once the forest was felled and we had made a clearing, a place where the light penetrated, a place where we could live and farm and feel safe. For thousands of years, our dream was to tame the forest, to control it.

We have largely succeeded. Today, the bulk of Norwegian forest is harvested through clear-cutting, in rational, efficient operations. And at the same time, as the forest becomes ever smaller and the population ever bigger, a different dream has emerged. The dream of finding our way back to the wilderness. To the old, untouched natural forest – to our origin and our identity.

And just as this kind of forest can stir me with an intense joy, I may without warning feel a dark claw trap this brightness in its grip, making me catch my breath abruptly. I can feel a profound sorrow because such forest barely exists any more. Because the few, tiny remaining fragments are under constant threat from economic growth, the demand for more resources, our notions of progress. Some people call this 'ecological grief' – an acknowledgement of how comprehensively we have altered the planet we live on.

Many people have favourite places in nature, spots they are fond of that form part of their identity. It is vital to preserve these. To lose a place like that is to lose a little of oneself.

Homo Indoorus – *Nature and Health*

Let's say you live to 100. If you are an average European, you will have spent 90 of those years indoors. Ever since we humans built our first, primitive houses 300,000 years ago, we have produced ever-growing swathes of indoor space. The indoor floorage of Manhattan, for example, is now three times larger than the surface area of the island itself.

The word ecology comes from *oikos*, meaning 'home'. And ecology is, precisely, the study of our home. We're not talking about living-room design, though, or the latest trend in kitchen styling, but nature, that teeming green (sometimes white) outside our windows. What do you really know about your own home – about nature? British scientists tested this. It turned out that half of 2,000 adult Britons were incapable of recognising a sparrow. In another survey, children were shown picture cards depicting common British plants and animals together with cards bearing

Pokémon figures. Eight- to eleven-year-olds were far better at identifying Pokémon 'species' than real species – like oak trees or badgers. Around 80 per cent of the Japanese fantasy figures were correctly named versus less than half of the real species.

Perhaps that should come as no surprise considering the changes in children's outdoor life and outdoor play. The point is illustrated by a 2008 article from British tabloid the *Daily Mail*, which looked at the changes in children's right to roam over four generations of eight-year-olds. Great-grandfather George was born in 1926 and was eight in 1934. His family lived in cramped conditions and George spent most of his free time outdoors, with no organised activity and no adult supervision. He often went to his favourite fishing pond, almost 10 kilometres away from home. In the next generation down, Grandfather Jack was eight in 1950. He was allowed to go and play in the local woods, a few kilometres away. He also walked to school alone every day. When Mum Vicky was eight in 1979, she played in the park and the neighbourhood where she lived; at a pinch, she could walk to the local swimming pool 800 metres away. Today, her son Edward plays in the garden. He's not allowed to walk to school – his mother drives him there. If he wants to cycle, his mother puts his bike in the back of the car and drives him to a place where they can cycle safely together.

And things are no better at kindergarten. A comparison of 200 new and older kindergartens in Oslo shows that the area per child has fallen by almost 13 square metres (nursery schools built before 1975 versus after 2006). Some 54 per cent of the area lost is the children's outdoor space, while the parking space and reception area has shrunk by just 2 per cent. This is largely explained by the fact that, whereas parking space is regulated by law, the requirement for a minimum-sized play area per child was, ironically, removed in 2006.

We adults also spend less time outside than before. We have become keyboard-clicking office drones by day, screen-fixated couch potatoes by night and at weekends. Does this mean that we have abandoned our original *oikos*, nature, and moved indoors, amid the parquet and pleated blinds? Most of us live our indoor life without thinking much about what we are missing out on. But our new lives as *Homo indoorus* have important consequences. The absence of nature can make us ill – and several mechanisms come into play here.

One thing is that regular contact with nature, in the form of soil, plants and animals, helps build our immune systems. This connection, known as the biodiversity hypothesis of health, involves an apparent link between the loss of biological diversity and the increasing incidence of non-infectious chronic disease. The diseases in question are the ones that make our immune systems run riot – multiple sclerosis (MS), rheumatoid arthritis, asthma, allergies, coeliac disease, inflammatory bowel diseases, type 1 diabetes.

Within this panorama, microorganisms play a key role. The loss of biological diversity is not just about passenger pigeons and rhinoceroses. It's also about microbes on and in our bodies – because each of us is a walking zoo, housing billions of bacteria. New estimates suggest that microbes account for around 200 grams of the weight of an average man. If we encounter fewer microbes in our lifetime because we're insufficiently exposed to soil, animals and green nature, that leaves our immune system less robust, making it easier for us to fall ill. New research in support of the biodiversity hypothesis is now coming thick and fast.

An entirely different, also growing, body of research explores the link between nature and mental health. The statistics show that on average around a quarter of all adult Norwegians have had mental health problems each year. Nature has solutions in store: keeping active in green surroundings can be a simple but effective remedy. It is good for your health and therefore good for society.

Many people have now heard the Japanese term *shinrin-yoku*, or forest bathing. The term first appeared in scientific literature in 1998, in a study showing that a stroll reduced blood sugar levels in diabetes patients. When I look it up now, I get more than 100 hits on Web of Science, an internet database of academic articles (and more than a million hits on Google). Studies here show that forest visits are good for everything from brain activity, stress hormones, pulse and blood pressure to self-reported mood, sleep and concentration.

And it doesn't necessarily have to be forest: other types of nature will do. A summary from 2019 concludes that there is 'strong proof of a connection between exposure to nature and health', although it adds that we still lack much of the knowledge needed to understand the cause of these effects. But since taking a stroll through nature is simple, free and has no side effects, there's not much to think through here: it's just a matter of tying our shoelaces and getting outside. Like the COVID-19 crisis made lots of Norwegians do – there seemed to be hikers everywhere and a hammock in every second tree in the forest around Oslo in the spring of 2020.

One last point is that we need to understand nature in order to want to look after it. Contact with nature and positive experience of nature in childhood make it more likely that you will be concerned about environmental issues as an adult. The fact that adults who are important to you spend time outdoors with you, showing you nature and helping you understand that nature means a lot to them, also plays a big role. So, we need to get outdoors. And look and touch and listen and smell. Breathe and taste and feel.

Savour the joy of just being outside, in a local wood, at the park, by the sea or on a wintry mountain. After all, how can we take care of nature if we're not familiar with it? How can we expect our children to do a better job of saving the climate and nature than we have if no one has taught them to love their fellow creatures?

Greenwashing, Whitewashing – Ornamental Lawns and Wild Gardens

Why are we so fond of certain types of groomed, manmade nature, like lawns? And how does knowledge of biology affect our views of what we think is ugly or pretty? A few years back, I travelled through California in the height of summer. In Central Valley, acre after acre is planted with almond trees, an extremely water-intensive species. There was a drought, and water restrictions were in place. In the city suburbs, the small front lawns bore withered, scorched witness to the gravity of the situation, but – in typical American style – someone had seen a business opportunity even here. A firm put up huge yard signs advertising its services: 'Got dead grass? Got a brown lawn? Paint it!' followed by a phone number.

This is where humankind's dream of the wilderness ends, I thought – as I surreptitiously snapped the sign through the car window. Spraying the withered tufts of grass in your

248

garden with green paint to fake living, local nature struck me as so utterly disheartening. As if it isn't bad enough having a real lawn in the first place.

Lawns are like green tarmac, with extremely limited biological diversity. Especially when we – in our mono-maniacal zeal to create 'perfect' lawns without a single flower – resort to all kinds of pesticides. In the US, where lawns cover an area equivalent in size to half of Norway, 34,000 tonnes of pesticide per year are applied to lawns alone. That diminishes and alters the natural soil fauna that would normally take on the task of recycling dead plants into new nutrients. As a result, Americans have to apply 41,000 tonnes of synthetic lawn fertiliser on top of that.

Why in the world do we do it? Why do we think lawns are so lovely? Why not have a flourishing, brightly coloured flower meadow, burgeoning with scents and sounds and insect life? Lawns are a cultural phenomenon that first emerged as a decorative element in French and English landscape gardens. Over the Renaissance it became a status symbol for aristocrats and the nobility to display their wealth by letting large areas of grass lie idle, as decoration, without any grazing animals. Could that help explain the lawn's popularity and its absolute dominance of green areas all over the world?

Nowadays, lawns account for the majority of green spaces in the world's cities, close to 70 per cent in some cases. In Sweden, the land area covered by lawn has doubled in 50 years. At the same time, the area given over to natural, flower-rich meadows has shrunk dramatically over the past 150 years. Meadows have been built over, or have become overgrown, turning into forest. That also applies to Norway. In the Oslo Fjord area alone, around half of the flower-rich spaces have vanished since the 1950s.

We spend so much time in neat, carefully planned park-style nature that we are becoming squeamish and delicate, and need to be warned about perfectly normal phenomena in nature. On my way into Germany's Hasbruch nature reserve a few years back, I encountered a big red sign. The local authorities explained that in a protected natural forest, like the one I was now venturing into, dead branches were not removed from the trees so it was conceivable that branches like these could fall down and, in the worst case, they might happen to hit me on the head (this accompanied by a dramatic illustration of a man collapsing beneath one of those dangerous falling branches). I was entering the forest at my own risk, it continued.

But there are alternatives to the aesthetic of tidiness, where pretty does not equate to uniform and natural variation is not conveyed as dangerous. By talking instead about the advantages of letting a park or garden or forest grow a little wilder, we can reverse the trend. A flower meadow with a profusion of colours and scents and buzzing life will bring pleasure to two- and six-legged species alike. Standing dead trees are nature's very own insect hotels and can house thousands of species, whereas an artificial, shop-bought insect hotel can house a dozen at best. A beautiful messy corner in the garden of a villa provides space for a variety of predatory insects, maybe even a hedgehog, and makes a significant contribution to improving our cities' biodiversity.

Smart as a Plant – Other Species Can Do More Than You Think

It's not all that easy to see the world from a non-human perspective. To imagine senses we don't have. To reflect on how the world would look if you were a tomato plant or a mimosa, for example. We are trapped by the limitations of our perceptions, often exacerbated by an arrogant notion that our way of dealing with life's challenges is the only or the best way, so we are constantly taken by surprise – like when we discover that a plant can respond to the sound of buzzing insects in less than three minutes, increasing the sugar content of its nectar to attract more pollinators. But plants are more like us than you may realise.

All living organisms have certain fundamental processes in common: acquiring food and energy, growing, secreting waste, moving and reproducing. All living things must also be capable of sensing and responding to their surroundings. That applies to plants too. It is obvious that plants sense gravity, because roots grow downwards and stems upwards. But plants have many more senses, despite their lack of specialised organs like eyes, ears or noses.

The discussion as to whether plants can hear, see, smell and feel got off on the wrong foot when a book published in the 1970s claimed that they grew quicker if you played classical music to them. As no proof was ever given for the assertion, this myth tarnished the reputation of all other research into plants' senses for a long time. If you look up 'plant perception' on Wikipedia, you will still find two different references: one based on physiological knowledge, the other on pseudo-scientific claims.

Now, there is little real point researching plants' musical tastes because neither Mozart nor Metallica have any ecological relevance whatsoever if you happen to be a dandelion. However, a great deal of research that has appeared in recent years supports the idea that plants are like your children and mine: they hear what they want to. The thale cress, for example, can distinguish between the noise of the leaf-chomping caterpillars of the green-veined white butterfly and the sound of wind or insect song: plants that have heard the ominous sound of the caterpillar will produce more defensive substances when they themselves get chewed later. A plant in the evening primrose family turns out to be able to perceive the sound of a buzzing bee (or a synthetic sound in the same frequency zone) and responds by producing sweeter nectar. How can the plant hear? We still don't understand the details, but it appears that the flower itself acts as a kind of outer ear. If the petals are removed, the response disappears too. We can only begin to speculate about the consequences this has for plants growing in today's noisy cities.

It is easier to accept that plants can see, in the sense that they react to light, especially red and blue light. This should be self-evident, since they need light to make sugar – so light means food if you are a plant. We've all seen the way plant shoots stretch towards the light source. This is because the photosensitive receptors in the upper part of the shoot send signals that make the cells on the shoot's shadow side stretch and grow longer, causing the plant to bend towards the light. Plants can also 'see' their leafy neighbours because the relationship between different wavelengths of red light alters when light is filtered through or reflected off other plants.

Smell, or the capacity to perceive chemical compounds in gaseous form, is also important for plants. The reason you should avoid storing apples on the kitchen counter with other fruits is that they secrete large amounts of ethylene, a substance that speeds up the ripening process. In nature, many types of fruit react to this olfactory substance from neighbouring fruit by producing their own ethylene. In this way, plants ensure the coordinated ripening of their fruit, which is an important way of attracting creatures that will eat the fruit and spread the seed. On your kitchen counter, however, the result is that the fruit lying near the apples quickly becomes overripe. You can test this out by placing two unripe bananas in separate zip-lock bags and adding an apple to one of them. The banana in this bag will ripen more quickly.

But plants also use smell in other ways. A classic study shows how parasitic American climbing plants called dodders pick up the aroma substances of neighbouring plants and stretch their swaying tendrils directly towards the unfortunate victim. Dodders can distinguish between the aromas of tomato, their favourite plant, and wheat, which they aren't so keen on. If you watch a speeded-up video of this, you'll realise that there's a lot of truth in the joking claim that plants are just slow animals.

It has also been shown that tomato plants, for example, can perceive the scents their neighbours secrete when caterpillars sink their teeth into their leaves, and react to these aromas by stepping up their own production of defensive substances. This leaves them better equipped when the caterpillars reach them. There are important nuances in the way the phenomenon is referred to: reacting to the aroma from the neighbouring plant is not the same as saying that the plant under attack 'warns' its fellow species – the first is logical in an evolutionary context, the second implies a conscious desire for communication that is unsupported by evidence.

Taste is intimately linked to smell: just think how flavourless food becomes when you have a blocked nose and can't smell. For plants too this is a fluid boundary because the same substances can be sensed in both gaseous form (as smell) and dissolved in water (when we refer to it as taste), so we can call it taste when plant roots perceive and react to chemical substances in the soil. Plants can use this to grow towards the spot where there is most water or food, or to recognise other plants' roots.

The best examples of plants' capacity to feel are found among insectivorous plants that close around their unfortunate prey – and mimosa too. Mimosa, also known as the sensitive or touch-me-not plant, is great fun. As its common names suggest, the leaves react to touch, as a defence against grazing. I remember an encounter with a mimosa on a forest hike in the tropics with small children as one of the occasions when being a biologist mum was a recipe for success: my three-year-old in particular couldn't get enough of stroking the tiny leaves with a podgy index finger and seeing them snap shut.

The mimosa is one of the plants that has been used in revolutionary and still-controversial experiments whose results suggest that plants can both learn and remember. When a mimosa is dropped repeatedly – like bungee jumping for flowers – it apparently gets used to the treatment and stops closing its leaves in response, even though it is still capable of doing so when subjected to other kinds of stress. As if that wasn't enough, the plant also remembers its tolerance for bungee jumping for a whole month.

Ideas like these are far from new. The English naturalist Charles Darwin wrote about plants' sensory perception and was of the opinion that a plant's radicles or root tips were not unlike 'the brain of one of the lower animals'. His son, botanist Francis Darwin, gave a lecture on the subject to the British Association for the Advancement of Science in 1908. According to the *New York Times*, which covered the 'vegetable psychology' story in a full-page spread complete with pictures, these ideas occasioned considerable consternation among the bearded scientists present.

And that is the core of the problem, taking us back to where we started. We humans suffer from plant-blindness – an inability to *see* our chlorophyll-filled distant relatives, and an unwillingness to grasp what hides behind that green wall. Or, as a botanist so generously pointed out not long ago: we don't just suffer from plant blindness but from anything-other-than-vertebrates blindness. On a planet that is dominated by plants (80 per cent of all life by weight) and bugs (75 per cent of all known plant and animal species), should we humans really take pride in remaining so short-sighted and self-obsessed?

With a Little Help from my Friends –
An Intricate Interaction

When we apply the term 'intrinsic value' to nature, this means we think of it as having a value in its own right, without being useful. But what does this mean in more concrete terms? The provisions of Norway's Nature Diversity Act state: 'Acknowledging the intrinsic value of nature involves an acceptance that nature has ideal rights, i.e. protection against injury, including the notion that other life forms, irrespective of whether or not they are useful to humans, have a self-evident right to exist. Within this lies an element of respect for the interactions in nature, an interplay in which the biotic and abiotic combine to form the complex and "fine-mesh" tapestry that constitutes nature'.

Demanding words: woolly and perhaps not so easy to grasp. Talking about nature's intrinsic value simply isn't easy for us non-philosophers. The language treads a narrow path as it wends its way between the briar patch of alien-ating terminology on the one hand and the quicksand of cliché and banality on the other. We have terminology for ecology, economics, philosophy, but we struggle a bit to find everyday language that can express nature's underlying meaning; – in a way that does not frame us humans as recipients and nature as a service provider. Perhaps this is easier to illustrate through an example – one of the countless, ingenious interactions that bind together all this complex tapestry we call nature.

Sometimes you need a bit of help from your friends – if, say, you're a premature baby and have to start your life in an incubator. That's also how it is for some of the most beautiful flowers we know: orchids. Many orchid seeds are incredibly tiny, almost like motes of dust. This is because the seed doesn't contain its own packed lunch the way those of other plants do. It doesn't have any food reserve at all for the shoot to live off until it has put down roots. This makes the orchid seed totally reliant on help from kindly friends – in this case mycorrhizal fungi, which fit themselves like a glove around (and partly inside) the plants' roots. Most plants on Earth have this kind of fungus glove – or maybe we should call them toe-socks – on their roots once they have grown to full size.

The unusual thing about orchids and their fungal pals is that the relationship starts so early. The fungus packs the feather-light orchid seed into a kind of incubator made of soft mycorrhizal threads. And in this incubator the helpless little seed is supplied with food and water at the outset, until it has formed its own roots, stem and leaves. Only through this interaction can the seed develop into a beautiful full-grown orchid.

With 28,000 known species, the orchid family is the largest in the plant kingdom. In fact, every tenth plant species on the planet is an orchid. They are strange, stunningly beautiful and extremely varied. Perhaps you know them best from the florists, where you can buy a white or purple species from the distant tropics for 15 quid. Most orchids are, in fact, tropical, but even in Norway's barren land, a good 40 different species manage to find a foothold in sparse forests and on calcium-rich hillocks. Some are small and modest in all their green pallor – like dwarf rattlesnake plantain or eggleaf twayblade. Others are big and strikingly beautiful, in shades of yellow or reddish-pink, like the lady's slipper or the chalk fragrant orchid.

Their subterranean BFFs are, however, about as invisible as Frodo when he's wearing the Ring. They produce strange fruiting bodies (the fungus's 'flowers', which we see above the ground – like chanterelles), and DNA analyses of the soil are needed to coax them onto the species lists. But what they lack in physical charms, they make up for with genus names fit for bona fide princesses: *Tulasnella* and *Tomentella*.

All the world over, orchids are threatened by habitat destruction, climate change and our unscrupulous quest for their beauty. Experts who assessed the extinction risk of around 1,000 of the world's orchid species found that an insane 57 per cent ended up with the status of globally, often acutely, threatened. Perhaps some will say, so what? What's the point of 28,000 different orchid species? Do we even need them, if we can produce vanilla flavouring using a different process anyway (*see also page 121*) and can make ornamental plants out of plastic?

Or what about the intrinsic value of parasites – which are so vast in number, and can endanger us or make us sick? Are we capable of loving species that we find repulsive, whose lifestyles border on the grotesque? Like the tongue-eating louse, a tiny marine crustacean with the looks of a wood louse that parasitises species such as clownfish (remember the orange-and-white-striped eponymous hero in *Finding Nemo*?).

As a young male (all the young are, in fact, males) the tongue-eating louse enters a new fish through its gills. If there isn't already a female tongue-biter inside the fish's jaw, it changes sex, develops longer claws and grows from teensy-weensy to being the size of ... yep, a fish tongue. The long claws come in handy because the next stage in the lifecycle is that the female sinks her claws into Nemo's tongue, halting his blood supply, which causes the tissue to die and the tongue to fall off.

But hey, don't despair! The tongue-biter provides a replacement for the lost tongue: herself. She attaches herself firmly to the stump of the tongue with her legs and becomes the fish's new tongue. Like a living prosthetic limb, eyes peering out of the fish's jaw, she is a reminder to us all that the issue of intrinsic value is far from simple.

I find it exciting yet demanding to discuss the intrinsic value of nature and species. It's easy to agree with the principles contained in the statement: other organisms have a self-evident right to live their lives without any obligation to provide value creation or be useful to us. Yet however we approach the subject, we can't get away from the fact that we are humans, and all our knowledge, all our judgements about right and wrong, all our ethical principles are filtered through our perspective, limited by what a human can – and wishes to – perceive.

If I feel a deep respect, even reverence, for the complex interplay between fungi and orchids, and wish to assert that it is important for me to know that there are thousands of completely useless orchids in the rainforest, flowers I will never see – can I, then, disconnect that value from myself as an acting individual, a recipient? How far can we stretch an ecocentric view of nature, in which nature always stands at the centre? Is it perhaps naive to think that we should be capable of shaking off the principles of evolution and placing other species ahead of ourselves? When push comes to shove, won't we always, inevitably, choose ourselves and our nearest and dearest?

I will leave these lines of thought to the natural philosophers. And to my own campfire contemplations next time I find myself in an old forest leaning up against the trunk of a pine that germinated before Descartes was born.

Lost Wilderness and New Nature – The Way Forward

I went to the woods because I wished to live
deliberately, to front only the essential facts of life,
and see if I could not learn what it had to teach, and
not, when I came to die, discover that I had not lived.
I did not wish to live what was not life, living is so
dear; nor did I wish to practise resignation, unless it
was quite necessary. I wanted to live deep and suck
out all the marrow of life, to live so sturdily and
Spartan-like as to put to rout all that was not life, to
cut a broad swath and shave close, to drive life into a
corner, and reduce it to its lowest terms.

HENRY THOREAU

Walden or Life in the Woods, 1854

Imagine you're standing in a tropical forest in Hawaii, on the island of Oahu. Around you, everything is lush and humid and green. You can see the tree trunks striving to the heavens, eternally competing with each other to capture the life-giving rays of the sun. Leaves are busy photosynthesising and building biomass, carbon is being stored in stems and soil. Perhaps you'll catch a sickly stench of rotten leaves on the ground, where fungi and insects are running their caretaking company. Raindrops, the remnants of an afternoon shower, still drip from the tree canopies, trickle into the earth, are cleansed. You can hear the birds too, stealing around in the branches, eating ripe fruit and spreading seeds. You can literally see, smell and hear nature's goods and services all around you.

I bet you'd think it was wild and wonderful here, too. And perhaps you and I wouldn't even see it – but the fact is that the forest around you is no wilderness. The forest you see consists solely of introduced, alien tree and plant species. And pretty much every single bird you'll find in the forest was introduced: from nature's point of view they don't belong in Hawaii – we humans are the ones who brought all these plants and birds to the island.

What are we supposed to think about that? Some will say: the forest *works*, doesn't it? It produces ecosystem services like there's no tomorrow. Complex interactions have arisen among species that only recently encountered one another, here in the forest on Oahu. Nature responds as always to the changes we make, by being dynamic, adapting, evolving further. Is there necessarily such an enormous difference between *indigenous* and *introduced*? And can we really say that one forest is better than another – or are they simply different?

Others think we must turn the spotlight on what is vanishing. We have lost so much already: several hundred domestic species, most of them utterly unique, species that cannot be found anywhere else in the world, have died out in Hawaii. We have brutally pruned the tree of evolution; eliminated species that might have helped make the forest more flexible, more robust – in the face of a changed climate, for example. Perhaps this was where it once lived, the plant that could have become the new cancer medicine, or the insect that might have given us new antibiotics? We will never know.

If we wish to retain all future opportunities, we must secure as much biodiversity and as much of the nature that is relatively untouched by humans as possible because, as renowned American nature writer Aldo Leopold said, 'To keep every cog and wheel is the first precaution of intelligent tinkering' – in other words, save all the species.

You won't just find new nature set up by us humans in Hawaii, of course. A third of the planet's ice-free surface area is covered by such entirely new ecosystems, which have no natural parallel. At the same time, the last remnants of wilderness – defined as large areas of land or sea that are free of human impact – are vanishing. In just 16 years, between 1993 and 2009, a wilderness area larger than India (or two times the size of Alaska, if you like) was lost. More than 77 per cent of the land (excluding the Antarctic) and 87 per cent of the sea has been altered by human activity. Five countries – Australia, the US, Brazil, Russia and Canada – are responsible for 70 per cent of the planet's remaining wild areas on land and at sea. Norway comes sixth on this list because of its seas.

Within my professional field, conservation biology, debate has raged in recent years between biologists with conflicting views on wilderness and the new ecosystems that appear as a result of our impact on nature. At one end of the spectrum, we have what we could call the 'wilderness people', or 'traditional conservationists', who champion a view of nature in which nature stands at the centre and we humans are just one species among many. They profess the classic conservationist perspective that originated in nineteenth-century North America and was espoused by wilderness enthusiasts such as author Henry Thoreau (in *Dead Poets' Society* (1989) every meeting of the poetry club starts with the famous quotation from Thoreau's book *Walden*: 'I went to the woods because I wished to live deliberately, to front only the essential facts of life'.)

The wilderness people place great weight on protected areas as a tool. Their opponents claim that these traditional conservationists are setting nature above humanity – for example, by arguing in favour of throwing local populations out of newly established conservation areas, or creating strict reserves to which humans simply don't have access.

At the other end of the spectrum lie what we might call the 'welfare-first folk' (proponents of 'new conservation'). They believe wilderness is a naive dream and an inappropriate goal. Protecting untouched nature is a battle we lost long ago – it is a sad but unavoidable fact that there's not a patch of our planet that hasn't been affected by humans, directly or indirectly. The new conservation biologists think it is time to stop grieving over the lost wilderness dream, over extinct mammals and vanished great auks. Now we must wipe away our tears and concentrate instead on saving the remnants so that we can ensure wellbeing and fair distribution of goods and services for the generations to come.

The means to this end, according to these scientists, involves a far more pragmatic view of conservation, in which consideration for human beings and our welfare comes first. This does not mean that we don't need nature, because that is an inescapable fact. But it does imply a liberal attitude to moving species around between continents, wherever it best suits us, and implies a cost-benefit approach to the conservation of species. If it is an endless, expensive job to keep the house cats we brought with us from Europe away from a flightless New Zealand bird species, let's move the bird to a desolate cat-free Pacific Island where it has never lived. Possibly just let it die out if it has no significance for us, because we'd be better off spending the money available in a way that is more useful to us and nature instead.

These conflicting ideas about conservation have sparked heated debate in my academic field. Perhaps, in reality, it's not a question of these two extremes: the wilderness people versus the welfare-first folk. Most conservation biologists, myself included, believe we must find a set of realistic compromises between the naive-romantic and the hyper-pragmatic approaches. To practise conservation biology is to deal with a never-ending series of dilemmas.

We will have to make crucial choices along this axis over the decades to come as, with ever-increasing frequency, we find ourselves confronted by the results of human activity and our comprehensive tampering with nature. That is why I think we should talk more about these issues, well aware that there are no obvious right answers. The discussion is vital in itself, because it sharpens our thinking, makes us more conscious of value choices – our own and those of others. Whatever happens, we will not escape unscathed: the world has been irrevocably altered by us humans and we must find a way forward, together.

Afterword

Another world is not only possible, she is on her way.
On a quiet day, I can hear her breathing.

ARUNDHATI ROY

In the Sonoran Desert, Arizona (home to our superhero the
Gila monster, incidentally), there is a peculiar collection of
buildings. The glass and steel cupolas and domes in
assorted shapes and sizes are reminiscent of the childhood
home of *Star Wars* hero Luke Skywalker on the fictional
planet of Tatooine – supplemented by greenhouses from a
modern botanical garden. This dual association isn't
entirely off-target because this is Biosphere 2, which was
built as a complete miniature world for plants and animals
and people – to test whether we could replicate a habitable
closed ecosystem. Its aims included finding out whether we
could establish ourselves elsewhere in space.

To cut a long story short, it didn't go especially well. And the problems started pretty soon after four women and four men were incarcerated in there in 1991, in a kind of two-year, ecologically themed *Big Brother* experiment. Most of the vertebrates and pretty much all of the pollinating insects died very quickly. An ant species that had not been invited along but simply stowed away quickly came to dominate the crew-list of creepy-crawlies, along with cockroaches. Bindweed ran amok, blocking out the sunlight for other plants, including food crops. The eight biospherians were constantly hungry and in some cases ate the seeds they had brought in with them, which were intended for planting – in violation of the premise. Oxygen levels fell to such dangerously low levels that the research leaders twice had to break the seal to pump in fresh oxygen.

Why was the experiment called Biosphere 2? Because Biosphere 1 is the Earth, our home planet. The place where these life systems actually work, where the inconceivable masses of invisible species in nature are intertwined in a slow, dynamic interaction, and this is how they deliver the natural goods and services we humans need to live. Not just to survive – by the skin of our teeth – for two years, as the eight people in Biosphere 2 did, but to live for thousands and hundreds of thousands of years, until we achieve the pretty good life the majority of the world's population lives today. The Biosphere experiment illustrates what science shows: that intact, species-rich ecosystems are more robust and better at delivering the goods and services than impoverished, species-poor systems.

A lot has improved in this world: the majority of the world's population now lives in middle-class societies and the proportion of people living in extreme poverty has fallen from more than a third in 1990 to below a tenth today. Infant mortality has plummeted, the number of people dying of malaria has halved in just 15 years, life expectancy has more than doubled since 1900 and now stands at over 70 years. If I had published this book in around 1800, nine out of 10 people on this planet would not have understood a word of it. Today, the situation is reversed: nine out of 10 people can read.

But our numerousness and our way of life take a massive toll on nature: between my birth in 1966 and today, the number of people on this planet has doubled, and our consumption of natural resources has doubled since 1980 alone. Of the millions of species that help save your life, one in eight are under threat of extinction. And all this has repercussions for us. Almost a quarter of the Earth's surface has deteriorated and now produces less than it did before, and every year 10 per cent of the world's annual gross product is lost as a result of this deterioration in nature.

We may argue and disagree about many details within this picture: the best way to forecast the numbers or the most effective political measures for achieving a desired change. But fundamental logic dictates that constantly increasing resource use and eternal growth are impossible on a planet with finite resources. The IPBES puts it as clearly as it can be said: we must enact transformative social change. We must think innovatively and differently.

We can do this. We *have* to do this. The COVID-19 crisis has shown us that we are capable of implementing dramatic measures – and quickly at that – when we realise that a great deal is at stake. Countries can collaborate, the management of economies can be changed, scientists can share data in real time, people the world over can cooperate, changing their everyday lives – once we realise that this is what it will take to save the world as we know it and as we wish to keep it. This means that we must change our behaviour in our everyday lives, not to mention factoring in the environment when we cast our votes.

And to offer some perspective: the World Economic Forum issues an annual report detailing which threats will have the greatest impact on humanity over the coming decade. In the 2020 report, for the first time in history, all five issues at the top of the list are environment-related threats: extreme weather; failure of climate change mitigation and adaption; human-made environmental damage and disasters; biodiversity loss and collapse in ecosystems and natural disasters. The window of opportunity in which we can limit these threats is still open, but it is in the process of closing.

Yet I believe in hope. Not a naive, if-we-shut-our-eyes-tight-it'll-probably-pass hope – but a hope for action based on respect for life and love of everything we do not want to lose.

The original Norwegian title of this book, *På naturens skuldre*, meaning *On the Shoulders of Nature*, illustrates several points I am keen to communicate: the obvious yet overlooked fact that nature is what supports us; nature is the whole and entire basis of our wellbeing. Without nature to lift us, our civilisation will fall.

The image also says something about proportional size and reciprocity. The sum of other species and individuals is infinitely larger than our human population. Think how wonderful it is to sit on big shoulders when you are small, the way my grandfather who taught me about golden plovers and coltsfoot used to lift me up on his shoulders if the hike got a bit too long for short, childish legs. But you mustn't squeeze your bearer's neck too tightly because then they won't be able to breathe. And if from time to time you need to grab hold of their hair to keep your balance – well, you must do so gently, without pulling.

And what a wonderful view we get, sitting like that. Let us use our position as *Homo sapiens*, the clever human, sitting up there on nature's shoulders, to look ahead to the future that awaits, where our children and grandchildren will live; the future whose foundations we are laying through the actions we take today.

Thanks

I owe thanks to my Norwegian editor, Solveig Øye, and all the other supportive people at Kagge Forlag, to my daughter Tuva Sverdrup-Thygeson for feedback on early drafts, and to co-editor and foreign agent Hans Petter Bakketeig at Stilton Literary Agency for discussions and important input while writing this book. Also, I offer my heartfelt gratitude to Lydia Good and Joel Simons, my encouraging editors at HarperCollins UK, and to my brilliant English translator Lucy Moffatt – as well as other translators and publishers, for making my writing accessible across the world.

English Common Names and
Their Latin Equivalents

albatross – Diomedeidae family
almond – *Prunus dulcis*
ant – Formicidae family
  Argentine – *Linepithema humile*
  fungus-farming – *Trachymyrmex turrifex*
aphid – Aphidoidea superfamily
apple – *Malus domestica*
aspen – some *Populus* spp.
auk, great – *Pinguinus impennis*
avocado – *Persea americana*

bamboo – Bambusoideae subfamily
bat – Chiroptera order
  fruit – Pteropodidae family
  Mexican free-tailed – *Tadarida brasiliensis*
beaver – *Castor* spp.
bee
  cuckoo bumblebee – *Psithyrus* spp.
  honeybee – *Apis* spp.
  orchid – Euglossini tribe
beech, European – *Fagus sylvatica*

beetle
  bark – Scolytinae subfamily
  cereal leaf – *Oulema melanopus*
  ground – Carabidae family
  hermit – *Osmoderma eremita*
  leaf – Chrysomelidae family
  longhorn – Cerambycidae family
  sexton – *Nicrophorus* spp.
  stag – Lucanidae family
beewolf – *Philanthus* spp.
birch – *Betula* spp.
  dwarf – *Betula nana*
bitterwood (family) – Simaroubaceae family
bitterwood (species) – *Quassia amara*
boar, wild – *Sus scrofa*
Brazil nut – *Bertholletia excelsa*
butterfly
  green-veined white – *Pieris napi*
  Morpho – *Morpho* spp.

cacao – *Theobroma cacao*
camel, 'yesterday's' – *Camelops hesternus*
caterpillar – Lepidoptera order
chicken – *Gallus gallus domesticus*
cicada – Cicadoidea superfamily
clove – *Syzygium aromaticum*
clover, red – *Trifolium pratense*
clownfish – Amphiprioninae subfamily
cockroach – in Blattodea order
cod – *Gadus* spp.
coffee – *Coffea arabica*
coltsfoot – *Tussilago farfara*
coral – Anthosoa class
corn – *Zea mays*

cotton grass – *Eriophorum* spp.
cotton plant – *Gossypium* spp.
cow – *Bos taurus*
cress, thale – *Arabidopsis thaliana*
crowberry – *Empetrum nigrum*

dandelion – *Taraxacum* spp.
deer
  Irish – *Megaloceros giganteus*
  red – *Cervus elaphus*
dodder – *Cuscuta* spp.
dog – *Canis familiaris*
dwarf rattlesnake plantain, *see under* orchid

eagle, golden – *Aquila chrysaetos*
earthworm – in Opisthopora order
eggleaf twayblade, *see under* orchid
elephant
  African savannah – *Loxodonta africana*
  dwarf – in Proboscidea order
evening primrose – Onagraceae family

fig tree – *Ficus* spp.
firefly – in Lampyridae family
flax – *Linum usitatissimum*
flea – Siphonaptera order
fly
  blow – Calliphoridae family
  deer – *Chrysops* spp.
  flesh – Sarcophagidae family
  flower/hover – Syrphidae family
  fruit – Drosophilidae family
  house – *Musca domestica*

fox
  Arctic – *Vulpes lagopus*
  red – *Vulpes vulpes*
foxglove – *Digitalis* spp.
frog, gastric-brooding – *Rheobatrachus* spp.
fungus
  birch polypore – *Fomitopsis betulina*
  honey – *Armillaria* spp.
  penicillin – *Penicillium* spp.
  red-belted conk – *Fomitopsis pinicola*
  tinder – *Fomes fomentarius*

Gila monster – *Heloderma suspectum*
giraffe – *Giraffa camelopardalis*
glow-worm, *see* firefly
goat – *Capra aegagrus hircus*
goose, Canada – *Branta canadensis*
grape – *Vitis* spp.
grass
  brome – *Bromus* sp.
  lyme – *Leymus arenarius*
  wavy hair – *Deschampsia flexuosa*

heather, common – *Calluna vulgaris*
hedgehog – Erinaceinae subfamily
hellebore – *Helleborus* spp.
hemp – *Cannabis sativa*
henbane – *Hyoscyamus niger*
hook-moss, floating – *Warnstorfia fluitans*
hornet, European – *Vespa crabro*
horseshoe crab – *Carcinoscorpius; rotundicauda; Limulus;*
  polyphemus; Tachypleus; gigas; Tachypleus tridentatus

indigo – *Indigofera tinctoria*

jellyfish, immortal – *Turritopsis dohrnii*

kingfisher – Alcedinidae family
knot, red – *Calidris canutus rufa*
knotweed, Japanese – *Reynoutria japonica*
krill – Euphausiacea order

lady's mantle – *Alchemilla* spp.
larch – *Larix* spp.
leech – Hirudinea subclass
lingonberry – *Vaccinium vitis-idaea*
lion, American cave – *Panthera atrox*
lotus, sacred – *Nelumbo nucifera*
louse, tongue-eating – *Cymothoa exigua*
lupine – *Lupinus* spp.
　　Alaskan lupine – *Lupinus nootkatensis*

magnolia – *Magnolia*
mallow – Malvaceae family
mammoth – *Mammuthus* spp.
mandrake – *Bryonia alba*; *Mandrogora* spp.
mangrove – *Rhizophora* spp.
mastodon – *Mammut* spp.
mimosa – *Mimosa* spp.
mistletoe, European – *Viscum album*
mosquito (malaria) – *Anopheles* spp.
mouse, deer – *Peromyscus* spp.
mugwort – some *Artemisia* spp.
mushroom
　　butter cap – *Rhodocollybia butyracea*
　　chanterelle – *Cantharellus* spp.
　　hedgehog – *Hydnum repandum*
　　porcini – *Boletus edulis*

mussel, freshwater pearl – *Margaritifera margaritifera*
myrrh – *Commiphora* spp.

oak – *Quercus* spp.
octopus – Octopoda order
opossum, Virginia – *Didelphis virginiana*
orchid – Orchidaceae family
   chalk fragrant – *Gymnadenia conopsea*
   dwarf rattlesnake plantain – *Goodyera repens*
   eggleaf twayblade – *Neottia ovata*
   lady's slipper – Cypripedioideae subfamily
   vanilla – *Vanilla planifolia*
owl – Strigiformes order
   boreal – *Aegolius funereus*

pangolin – *Manis*; *Phataginus*; *Smutsia* spp.
penguin – *Aptenodytes*; *Eudyptes*; *Eudyptula*; *Megadyptes*;
   Pygoscelis; *Spheniscus* spp.
perch – *Perca* spp.
pig – *Sus* spp.
pigeon – Columbidae family
   passenger – *Ectopistes migratorius*
pine – *Pinus* spp.
   lodgepole (twisted) – *Pinus contorta*
plane, London – *Platanus* × *acerifolia*
plover, golden – *Pluvialis apricaria*
poplar – some *Populus* spp.
poppy – *Papaver* spp.

raspberry – *Rubus idaeus*; *Rubus strigosus*
raven – *Corvus* spp.
redwood, coast – *Sequoia sempervirens*
redwood, giant – *see* sequoia, giant
reindeer – *Rangifer tarandus*

rhinoceros – *Ceratotherium*; *Dicerorhinus*; *Diceros*; *Rhinoceros* spp.
  woolly – *Coelodonta antiquitatis*
rice – *Oryza glaberrima*; *Oryza sativa*

sabre-toothed tiger – *Smilodon* spp.
salmon
  Atlantic – *Salmo salar*
  Pacific – *Oncorhynchus* spp.
saxifrage, purple – *Saxifraga oppositifolia*
scampi – *Nephrops norvegicus*
sea anemone – Actiniaria order
sequoia, giant – *Sequoiadendron giganteum*
shark – Selachimorpha superorder
sheep – *Ovis aries*
shrew – Soricidae family
slime mould, dog vomit – *Fuligo septica*
sloth, giant – *Megatherium americanum*
slug, Spanish – *Arion vulgaris*
soya – *Glycine max*
sparrow – *Passer* spp.
spruce
  Norway – *Picea abies*
  Sitka – *Picea sitchensis*
spurge, cushion – *Euphorbia epithymoides*
stingray – Myliobatoidei suborder
strawberry, wild – *Fragaria vesca*
swift
  chimney – *Chaetura pelagica*
  common – *Apus apus*
swordfish – *Xiphias gladius*

termite – Termitoidae epifamily
tick – Ixodida suborder
tiger – *Panthera tigris*

283

tomato – *Solanum lycopersicum*
turtle, pond – Emydidae family
tree of heaven – *Ailanthus altissima*

wasp, fig – Agaonidae family
water lily – Nymphaeaceae family
whale
  blue – *Balaenoptera musculus*
  humpback – *Megaptera novaeangliae*
  northern minke – *Balaenoptera acutorostrata*
  sperm – *Physeter macrocephalus*
wheat – *Tritium aestivum*
whipworm – *Trichuris trichiura*
willow – *Salix* spp.
wolf – *Canis lupus*
wolverine – *Gulo gulo*
woodpecker – Picidae family
worm, bone (zombie) – *Osedax* spp.
wormwood, sweet – *Artemisia annua*

yeast, brewer's – *Saccharomyces cerevisiae*
yew – *Taxus* spp.
  European – *Taxus baccata*
  Pacific – *Taxus brevifolia*

Sources

Epigraph

The quotation from Rachel Carson, in which she explains her motivation for writing *Silent Spring* (1962), comes from the collected letters in the book *Always, Rachel: The Letters of Rachel Carson and Dorothy Freeman, 1952–1964 – The Story of a Remarkable Friendship*, published in 1996.

Preface

Bar-On, Y.M. et al. 'The biomass distribution on Earth', *PNAS* 115: 6506–6511 (2018)

1. Water of Life

New York City: The Champagne of Drinking Water

Appleton, A.F. 'How New York city used an ecosystem services strategy' (2002) https://www.cbd.int/financial/pes/usa-pesnewyork.pdf

Hanlon, J.W. 'Watershed protection to secure ecosystem services', *The New York City Watershed Governance Arrangement* 1: 1–6 (2017)

Sagoff, M. 'On the value of natural ecosystems: The Catskills parable', *Politics and the Life Sciences* 21: 19–25 (2002)
On the new dispensation for federal regulations: https://www.nytimes.com/2018/01/18/nyregion/new-york-city-water-filtration.html

Freshwater Pearl Mussels – Caretakers of the Water System
Jakobsen, P. *Samlerapport om kultivering og utsetting av elvemusling 2018*, Universitetet i Bergen (read 2019)
Jakobsen, P. et al. *Rapport 2013 for prosjektet: Storskala kultivering av elvemusling som bevaringstiltak*, Universitetet i Bergen (red. 2014)
Larsen, B.M. Elvemusling (*Margaritifera margaritifera L.) Litteraturstudie med oppsummering av nasjonal og internasjonalkunnskapsstatus*, NINA-Fagrapport 28 (1997)
Larsen, B.M. *Handlingsplan for elvemusling (Margaritifera margaritifera L.) 2019–2028.* Miljødirektoratet Rapport M-1107 (2018)
Larsen, B.M. et al. *Overvåking av elvemusling i Norge. Årsrapport for 2018*, NINA Rapport 1686 (2019)
Lopes-Lima, M. et al. 'Conservation status of freshwater mussels in Europe: state of the art and future challenges', *Biol Rev Camb Philos Soc* 92: 572–607 (2016)
Vaughn, C.C. 'Ecosystem services provided by freshwater mussels', *Hydrobiologia* 810: 15–27 (2018)
About pearl-fishing in Norway: https://www.jaermuseet.no/samlingar/wp-content/uploads/sites/16/2011/06/2004.07-D%C3%A5-perlefangsten-i-H%C3%A5elva-var-kongeleg-privilegium-2.pdf

Poisoners and Purifying Moss
Gerhardt, K.E. et al. 'Opinion: Taking phytoremediation from proven technology to accepted practice', *Plant Science* 256: 170–185 (2017)

Sandhi, A. et al. 'Phytofiltration of arsenic by aquatic moss (*Warnstorfia fluitans*)', *Environmental Pollution* 237: 1098–1105 (2018)

Sophie Johannesdotter's execution: https://www.nb.no/items/ URN:NBN:no-nb_digavis_fredriksstadtilskuer_null_ null_18760219_12_21_1

Uppal, J.S. et al. 'Arsenic in drinking water – recent examples and updates from Southeast Asia', *Current Opinion in Environmental Science & Health* 7: 126–135 (2019)

2. A Gargantuan Grocery Store

The quotation about food is from the Taittirīya Upanishads 10 III 6, drawn from *Sixty Upaniṣads of the Veda, Part 1*, by Paul Deussen and V.M. Bedekar (1980)

Something Brewing – Wasps and Wine

About the Official Microbe of the state of Oregon: https://gov. oregonlive.com/bill/2013/HCR12/

McGovern, P.E. et al. 'Fermented beverages of pre- and protohistoric China', *PNAS* 101: 17593–17598 (2004)

Stefanini, I. et al. 'Role of social wasps in *Saccharomyces cerevisiae* ecology and evolution', *PNAS* 109: 13398 (2012)

Ibid. 'Social wasps are a *Saccharomyces* mating nest', *PNAS* 113: 2247 (2016)

Statistics from http://www.fao.org/statistics/en/

The Inger Hagerup quotation is from 'The Wasp' in *Little Parsley*, translated by Becky L. Crook and published by Enchanted Lion Books (2019)

If You Are What You Eat, You're Walking Grass

IPBES. Chapter 2.3. 'Status and Trends' – *NCP: The Global Assessment Report on BIODIVERSITY AND ECOSYSTEM SERVICES* (draft) (2019)

Milesi, C. et al. 'Mapping and modeling the biogeochemical cycling of turf grasses in the United States', *Environ Manage* 36:426–38 (2005)

The poem 'Grass' by Joyce Sidman is from *Ubiquitous: Celebrating Nature's Survivors*, Houghton Mifflin Harcourt, 2010

An Avalanche of Extinctions – The Megafauna that Vanished

About the avocado: https://www.smithsonianmag.com/ arts-culture/why-the-avocado-should-have-gone-the-way-of-the-dodo-4976527/

Doughty, C.E. et al. 'The impact of the megafauna extinctions on savanna woody cover in South America', *Ecography* 39: 213–222 (2016)

Faurby, S. et al. 'Historic and prehistoric human-driven extinctions have reshaped global mammal diversity patterns', *Diversity and Distributions* 21: 1155–1166 (2015)

Galetti, M. et al. 'Ecological and evolutionary legacy of megafauna extinctions', *Biological Reviews* 93: 845–862 (2018)

Janzen, D.H. et al. 'Neotropical anachronisms: the fruits gomphotheres ate', *Science* 215: 19–27 (1982)

Keesing, F. et al. 'Cascading consequences of the loss of large mammals in an African savanna', *BioScience* 64: 487–495 (2014)

Malhi, Y. et al. 'Megafauna and ecosystem function from the Pleistocene to the Anthropocene', *PNAS* 113: 838–846 (2016)

Pires, M.M. et al. 'Reconstructing past ecological networks: the reconfiguration of seed-dispersal interactions after megafaunal extinction', *Oecologia* 175: 1247–1256 (2014)

Sandom, C. et al. 'Global late Quaternary megafauna extinctions linked to humans, not climate change', *Proc. Royal Soc.* B: 281: 20133254 (2014)

Smith, F.A. et al. 'Megafauna in the Earth system', *Ecography* 39: 99–108 (2016) This is the source of the quotation: 'Only recently have we begun to appreciate ...'

Smith, F.A. et al. 'Body size downgrading of mammals over the late Quaternary', *Science* 360: 310–313 (2018)

Steadman, D.W. et al. 'Asynchronous extinction of late Quaternary sloths on continents and islands', *PNAS*, 102:11763–11768 (2005)

Surovell, T. et al. 'Global archaeological evidence for proboscidean overkill', *PNAS* 102: 6231–6236 (2005)

Van Der Geer, A.A.E. et al. 'The effect of area and isolation on insular dwarf proboscideans', *Journal of Biogeography*, 43:1656–1666 (2016)

Meat-hungry – Past and Present

Bar-On, Y.M. et al. 'The biomass distribution on Earth', *PNAS* 115: 6506–6511 (2018)

Chaboo, C. et al. 'Beetle and plant arrow poisons of the Ju|'hoan and Hai||om San peoples of Namibia (*Insecta, Coleoptera, Chrysomelidae; Plantae, Anacardiaceae, Apocynaceae, Burseraceae*)', *Zookeys* 558: 9–54 (2016)

Statistics drawn from https://ourworldindata.org/meat-production and https://www.nationalgeographic.com/what-the-world-eats/

The Sea – The Last Healthy Part of a Sick World?

FAO. *The State of World Fisheries and Aquaculture* 2018 (2018)

FAO. *FAO Yearbook. Fishery and Aquaculture Statistics* 2017 (2019)

Pauly, D. et al. 'Fishing down marine food webs', *Science* 279:860 (1998)

The quotation 'It is untrue that the Sea is faithless ...' is from *Garman & Worse* by Alexander L. Kielland (1880)

Thurstan, R.H. et al. 'The effects of 118 years of industrial fishing on UK bottom trawl fisheries', *Nature Communications* 1: 1–6 (2010). Also http://www.fao.org/fishery/static/Yearbook/YB2017_USBcard/root/aquaculture/yearbook_aquaculture.pdf

Shifting Baseline: Why We Don't Notice Deterioration

McClenachan, L. 'Documenting loss of large trophy fish from the Florida Keys with historical photographs', *Conservation Biology* 23:636–643 (2009)

Pauly, D. et al. 'Fishing down marine food webs', *Science* 279:860 (1998)

The excerpt of a poem is from 'Metamorphosis' by Anja Konig, from the collection *Animal Experiments*, Bad Betty Press, 2020

3. The World's Biggest Buzz

The Blossoms and the Bees

Biesmeijer, J.C. et al. 'Parallel declines in pollinators and insect-pollinated plants in Britain and the Netherlands', *Science* 313:351–354 (2006)

Carvalheiro, L.G. et al. 'Species richness declines and biotic homogenisation have slowed down for NW-European pollinators and plants', *Ecology Letters* 16: 870–878 (2013)

Garibaldi, L.A. et al. 'Wild pollinators enhance fruit set of crops regardless of honey bee abundance', *Science* 339: 1608–1611 (2013)

Hallmann, C.A. et al. 'More than 75 percent decline over 27 years in total flying insect biomass in protected areas', *PLOS ONE* 12:e0185809 (2017)

IPBES. *The Global Assessment Report on Biodiversity and Ecosystem Services*. Complete draft version (2019)

Klein, A.-M. et al. 'Importance of pollinators in changing landscapes for world crops', *Proceedings of the Royal Society B: Biological Sciences* 274: 303–313 (2007)

Lister, B.C. et al. 'Climate-driven declines in arthropod abundance restructure a rainforest food web', *PNAS* 115:E10397-E10406 (2018)

Mallinger, R.E. et al. 'Species richness of wild bees, but not the use of managed honeybees, increases fruit set of a pollinator-dependent crop', *Journal of Applied Ecology* 52: 323–330 (2015)

Piotrowska, K. 'Pollen production in selected species of anemophilous plants', *Acta Agrobotanica* 61: 41–52 (2012)

Potts, S.G. et al. 'Global pollinator declines: trends, impacts and drivers', *Trends in Ecology & Evolution* 25: 345–353 (2010)

Powney, G.D. et al. 'Widespread losses of pollinating insects in Britain', *Nature Communications* 10: 1018 (2019)

Rader, R. et al. 'Non-bee insects are important contributors to global crop pollination', *PNAS* 113: 146–151 (2016)

Sánchez-Bayo, F. et al. 'Worldwide decline of the entomofauna: A review of its drivers', *Biological Conservation* 232: 8–27 (2019)

Seibold, S. et al. 'Arthropod decline in grasslands and forests is associated with landscape-level drivers', *Nature* 574: 671–674 (2019)

van Klink, R. et al. 'Meta-analysis reveals declines in terrestrial but increases in freshwater insect abundances', *Science* 368: 417 (2020)

Blue Honey Makes Beekeepers See Red

http://honeycouncil.ca/archive/chc_poundofhoney.php

https://www.reuters.com/article/us-france-bees/blue-and-green-honey-makes-french-beekeepers-see-red-idUSBRE8930MQ20121004

Two Flower Flies with One Swat

Dunn, L. et al. 'Dual ecosystem services of syrphid flies (*Diptera: Syrphidae*): pollinators and biological control agents', *Pest Management Science* 76: 1973–1979 (2020)

Hu, G. et al. 'Mass seasonal bioflows of high-flying insect migrants', *Science* 354: 1584–1587 (2016)

Lázaro, A. et al. 'The relationships between floral traits and specificity of pollination systems in three Scandinavian plant communities', *Oecologia* 157: 249–257 (2008)

Maier, C.T. et al. 'Dual mate-seeking strategies in male syrphid flies (*Diptera: Syrphidae*)', *Annals of the Entomological Society of America*, 72: 54–61 (1979)

Wotton, K.R. et al. 'Mass seasonal migrations of hoverflies provide extensive pollination and crop protection services', *Current Biology* 29: 2167–2173.e5 (2019)

Brazil Nuts and Flying Perfume Flacons

Dressler, R.L. 'Biology of the orchid bees (*Euglossini*)', *Annual Review of Ecology Evolution, and Systematics*. 13: 373–394 (1982)

Humboldt, A. et al. *Personal Narrative of Travels to the Equinoctial Regions of America, During the Year 1799–1804 – Volume 2,* George Bell & Sons, 1907

Maues, M. 'Reproductive phenology and pollination of the Brazil nut tree (*Bertholletia excelsa*) in eastern Amazonia', (1998)

Peres, C.A. 'Demographic threats to the sustainability of Brazil nut exploitation', *Science* 302: 2112–2114 (2003)

Photo showing how beautiful orchid bees are: http://gilwizen.com/orchidbees/

Sazima, M. et al. 'The perfume flowers of *Cyphomandra* (*Solanaceae*): Pollination by euglossine bees, bellows mechanism, osmophores, and volatiles', *Plant Systematics and Evolution* 187:51–88 (1993)

The Fig Tree and the Fig Wasp: Millions of Years of Loyalty and Treachery

Barling, N. et al. 'A new parasitoid wasp (*Hymenoptera: Chalcidoidea*) from the Lower Cretaceous Crato Formation of Brazil: The first Mesozoic Pteromalidae', *Cretaceous Research* 45:258–264 (2013)

Compton, S.G. et al. 'Ancient fig wasps indicate at least 34 Myr of stasis in their mutualism with fig trees', *Biology* Letters 6:838–842 (2010)

Denham, T. 'Early fig domestication, or gathering of wild parthenocarpic figs?' *Antiquity* 81: 457–461 (2007)

Hossaert-McKey, M. et al. 'How to be a dioecious fig: Chemical mimicry between sexes matters only when both sexes flower synchronously', *Scientific Reports* 6: 21236 (2016)

Janzen, D.H. 'How to be a fig', *Annual Review of Ecology and Systematics* 10: 13–51 (1979)

Kuaraksa, C. et al. 'The use of Asian ficus species for restoring tropical forest ecosystems', *Restoration Ecology* 21: 86–95 (2013)

Shanahan, M. et al. 'Fig-eating by vertebrate frugivores: a global review', *Biological Reviews, Cambridge Philosophical Society* 76: 529–572 (2001)

Thornton, I., W.B. et al. 'The role of animals in the colonization of the Krakatau Islands by fig trees (*Ficus species*)', *Journal of Biogeography* 23: 577–592 (1996)

Zahawi, R.A. et al. 'Tropical secondary forest enrichment using giant stakes of keystone figs', *Perspectives in Ecology and Conservation* 16: 133–138 (2018)

4. A Well-stocked Pharmacy

Alves, R.R. et al. 'Biodiversity, traditional medicine and public health: where do they meet?' *Journal of Ethnobiology and Ethnomedicine* 3: 14 (2007)

Calixto, J.B. 'The role of natural products in modern drug discovery', *Anais da Academia Brasileira de Ciências* 91 (2019)

Pharmaceuticals sector revenues: https://www.statista.com/topics/1764/global-pharmaceutical-industry/

Wormwood Versus Malaria

Cachet, N. et al. 'Antimalarial activity of simalikalactone E, a new quassinoid from *Quassia amara L.* (*Simaroubaceae*)', *Antimicrobial Agents and Chemotherapy* 53: 4393–4398 (2009)

Carter, G.T. 'Natural products and Pharma 2011: strategic changes spur new opportunities', *Natural Product Reports* 28: 1783–1789 (2011)

Gavin, M.C. 'Conservation implications of rainforest use patterns: mature forests provide more resources but secondary forests supply more medicine', *Journal of Applied Ecology* 46:1275–1282 (2009)

Kung, S.H. et al. 'Approaches and recent developments for the commercial production of semi-synthetic artemisinin', *Frontiers in Plant Science* 9: 87–87 (2018)

Newman, D.J. et al. 'Natural products as sources of new drugs over the 30 years from 1981 to 2010', *Journal of Natural Products* 75: 311–335 (2012)

Su, X.-Z. et al. 'The discovery of artemisinin and the Nobel Prize in Physiology or Medicine. Science China', *Life Sciences* 58:1175–1179 (2015)

The French biopiracy case: https://www.sciencemag.org/news/2016/02/french-institute-agrees-share-patent-benefits-after-biopiracy-accusations

Vigneron, M. et al. 'Antimalarial remedies in French Guiana: A knowledge attitudes and practices study', *Journal of Ethnopharmacology* 98: 351–360 (2005)

A Messenger Bearing Medicinal Mushrooms

Capasso, L. '5300 years ago, the Ice Man used natural laxatives and antibiotics', *The Lancet* 352: 1864 (1998)

Hassan, M.M. et al. 'Cyclosporin', *Analytical Profiles of Drug Substances* 16: 145–206 (1987)

Pleszczyńska, M. et al. '*Fomitopsis betulina* (formerly *Piptoporus betulinus*): the Iceman's polypore fungus with modern biotechnological potential', *World Journal of Microbiology and Biotechnology* 33: 83 (2017)

Yew's Whispered Wisdom

Allington-Jones, L. 'The Clacton spear: The last one hundred years', *Archaeological Journal* 172: 273–296 (2015)

Holtan, D. '*Barlinda Taxus baccata L. i Møre og Romsdal – på veg ut?*' *Blyttia* 59: 197–205 (2001)

Lines from 'Ash Wednesday', T.S. Eliot, Faber & Faber, originally published in 1930

Minke whaling with yew bows: https://www.kyst-norge. no/?k=2909&id=16004&aid=8396&daid=2604

Old yew in Scotland: https://www.woodlandtrust.org.uk/ blog/2018/01/ancient-yew-trees/

One of the best plant-based cancer treatments available: https:// www.cancer.gov/research/progress/discovery/taxol

Paclitaxel market revenue: https://www.reportsweb.com/reports/ global-paclitaxel-market-growth-2019-2024

Rao, K.V. 'Taxol and related taxanes. I. Taxanes of *Taxus brevifolia* bark', *Pharmaceutical Research* 10: 521–4 (1993)

Suffness, M. *Taxol: Science and Applications*, CRC Press, Boca Raton, FL, red. 1995

Żwawiak, J. et al. 'A brief history of taxol', *Journal of Medical Sciences* 1: 47 (2014)

Monster Spit Slays Diabetes

Background, scientist: https://www.nia.nih.gov/news/ exendin-4-lizard-laboratory-and-beyond and https://www. goldengooseaward.org/awardees/diabetes-medication

DeFronzo, R.A. et al. 'Effects of exenatide (exendin-4) on glycemic control and weight over 30 weeks in metformin-treated patients with type 2 diabetes', *Diabetes Care* 28: 1092–1100 (2005)

Drucker, D.J. et al. 'The incretin system: glucagon-like peptide-1 receptor agonists and dipeptidyl peptidase-4 inhibitors in type 2 diabetes', *Lancet* 368: 1696–1705 (2006)

Eng, J. et al. 'Isolation and characterization of exendin-4, an exendin-3 analog, from Heloderma-suspectum venom – further evidence for an exendin receptor on dispersed acini from guinea-pig pancreas', *Journal of Biological Chemistry* 267:7402–7405 (1992)

Exenatide, ranked no. 260 on the list of most commonly prescribed medicines in the US, with 1,635, 146 prescriptions in 2020: https://clincalc.com/DrugStats/Top300Drugs.aspx

Fedele, E. et al. 'Glucagon-like peptide 1, neuroprotection and neurodegenerative disorders', *Journal of Biomolecular Research & Therapeutics* 5 (2016)

Fry, B.G. et al. 'Early evolution of the venom system in lizards and snakes', *Nature* 439: 584–588 (2006)

Goke, R. et al. 'Exendin-4 is a high potency agonist and truncated exendin-(9-39)-amide an antagonist at the glucagon-like peptide 1-(7-36)-amide receptor of insulin-secreting beta-cells', *Journal of Biological Chemistry* 268: 19650–19655 (1993)

Grieco, M. et al. 'Glucagon-like peptide-1: A focus on neurodegenerative diseases', *Frontiers in Neuroscience* 13 (2019)

Holscher, C. 'Central effects of GLP-1: new opportunities for treatments of neurodegenerative diseases', *Journal of Endocrinology* 221: T31–T41 (2014)

Kamei, N. et al. 'Effective nose-to-brain delivery of exendin-4 via coadministration with cell-penetrating peptides for improving progressive cognitive dysfunction', *Scientific Reports* 8: 17641 (2018)

Meier, J.J. 'GLP-1 receptor agonists for individualized treatment of type 2 diabetes mellitus', *Nature Reviews Endocrinology* 8: 728–742 (2012)

Ohshima, R. et al. 'Age-related decrease in glucagon-like peptide-1 in mouse prefrontal cortex but not in hippocampus despite the preservation of its receptor', *American Journal of BioScience* 3: 11–27 (2015)

Strimple, P.D. et al. 'Report on envenomation by a Gila monster (*Heloderma suspectum*) with a discussion of venom apparatus, clinical findings, and treatment', *Wilderness & Environmental Medicine* 8:111–116 (1997)

The bite, 'like hot lava coursing through your veins', from: https://www.youtube.com/watch?v=swlozUKuvFI

The Giant Gila Monster film: see e.g. https://www.youtube.com/watch?v=Jdn-OCWEN00

Blue Blood Saves Lives

About the two scientists and the story of the discovery: https://www.goldengooseaward.org/awardees/horseshoe-crab-blood

Bolden, J. et al. 'Application of Recombinant Factor C reagent for the detection of bacterial endotoxins in pharmaceutical products', *PDA Journal of Pharmaceutical Science and Technology* 71: 405–412 (2017)

Ding, J.L. et al. 'A new era in pyrogen testing', *Trends in Biotechnology* 19: 277–81 (2001)

John, A. et al. 'A review on fisheries and conservation status of Asian horseshoe crabs', *Biodiversity and Conservation*: 1–26 (2018)

Maloney, T. et al. 'Saving the horseshoe crab: A synthetic alternative to horseshoe crab blood for endotoxin detection', *PLOS Biology* 16: e2006607 (2018)

Price of horseshoe crab blood: https://www.theguardian.com/environment/2018/nov/03/horseshoe-crab-population-at-risk-blood-big-pharma

Red knot sub-species (*Calidris canutus rufa*), dwindling population: https://fws.gov/northeast/red-knot/

rFC assay incorporated into the European pharmacopoeia: https://www.cleanroomtechnology.com/news/article_page/Recombinant_Factor_C_assay_to_aid_demand_for_LAL_endotoxin_testing/163099

Status of the four horseshoe crab species on the global red list: https://www.iucnredlist.org/search?query=Horseshoe%20Crab&searchType=species

Drugs from Bugs – Insects as a New Source of Antibiotics

Bibb, M.J. 'Understanding and manipulating antibiotic production in actinomycetes', *Biochemical Society Transactions* 41: 1355–64 (2013)

Cassini, A. et al. 'Attributable deaths and disability-adjusted life years caused by infections with antibiotic-resistant bacteria in the EU and the European Economic Area in 2015: A population-level modelling analysis', *The Lancet Infectious Diseases* 19:56–66 (2019)

Chevrette, M.G. et al. 'The antimicrobial potential of Streptomyces from insect microbiomes', *Nature Communications* 10: 516 (2019)

Costa-Neto, E.M. 'Entomotherapy, or the medicinal use of insects', *Journal of Ethnobiology* 25: 93–114 (2005)

Goettler, W. et al. 'Morphology and ultrastructure of a bacteria cultivation organ: The antennal glands of female European beewolves, *Philanthus triangulum* (*Hymenoptera, Crabronidae*)', *Arthropod Structure & Development* 36: 1–9 (2007)

Jühling, J. *Die Tiere in der deutschen Volksmedizin alter und neuer Zeit.* Polytechnische Buchhandlung (R. Schulze). Digitally available at https://dlcs.io/pdf/wellcome/pdf-item/b24856162/0#_ga=2.18265337.119250862.1579684184-1935579294.1579684184 (1900)

Kaltenpoth, M. et al. 'Symbiotic bacteria protect wasp larvae from fungal infestation', *Current Biology* 15: 475–479 (2005)

Kroiss, J. et al. 'Symbiotic Streptomycetes provide antibiotic combination prophylaxis for wasp offspring', *Nature Chemical Biology* 6:261–263 (2010)

Meyer-Rochow, V.B. 'Therapeutic arthropods and other, largely terrestrial, folk-medicinally important invertebrates: A comparative survey and review', *Journal of Ethnobiology and Ethnomedicine* 13: 9–9 (2017)

O'Neill, J. 'The review on antimicrobial resistance. tackling drug-resistant infections globally: Final report and recommendations'.

Available at: http://amr-review.org/sites/default/files/160518_
Final%20paper_with%20cover.pdf (2016)

Seabrooks, L. et al. 'Insects: An underrepresented resource for
the discovery of biologically active natural products', *Acta
Pharmaceutica Sinica* B 7: 409–426 (2017)

Strohm, E. et al. 'Leaving the cradle: How beewolves (*Philanthus
triangulum* F.) obtain the necessary spatial information for
emergence', *Zoology Jena* 98: 137–146 (1994/5)

When the Kids Make You Puke

Corben, C.J. et al. 'Gastric brooding: Unique form of parental care
in an Australian frog', *Science* 186: 946–947 (1974)

Fanning, J.C. et al. 'Converting a stomach to a uterus: The
microscopic structure of the stomach of the gastric brooding
frog *Rheobatrachus silus*', *Gastroenterology* 82: 62–70 (1982)

IPBES. *The Global Assessment Report on Biodiversity and
Ecosystem Services*. Complete draft version (2019)

Liem, D.S. 'A new genus of frog of the family *Leptodactylidae* from
south-east Queensland, Australia', *Memoirs of the Queensland
Museum*, 16(3), 459–470 (1973)

Mark, N.H. et al. 'Biochemical studies on the relationships of the
gastric-brooding frogs, genus *Rheobatrachus*', *Amphibia-Reptilia*
8: 1–11 (1987)

Red list status *Rheobatrachus silus* https://www.iucnredlist.org/
species/19475/8896430

Red list status *Rheobatrachus vitellinus*: https://www.iucnredlist.
org/species/19476/8897826

Reojas, C. 'The southern gastric-brooding frog', *The Embryo Project
Encyclopedia*. https://embryo.asu.edu/pages/southerngastric-
brooding-frog-0 (2019)

Scheele, B.C. et al. 'Amphibian fungal panzootic causes catastrophic
and ongoing loss of biodiversity', *Science* 363:1459–1463 (2019)

Tyler, M.J. et al. 'Oral birth of the young of the gastric brooding
frog *Rheobatrachus silus*', *Animal Behaviour* 29: 280–282 (1981)

Ibid. 'Inhibition of gastric acid secretion in the gastric brooding frog, *Rheobatrachus silus*', *Science* 220: 609–610 (1983)

Mini-jellyfish and the Mysteries of Immortality

Alves, C. et al. 'From marine origin to therapeutics: The antitumor potential of marine algae-derived compounds', *Frontiers in pharmacology* 9: 777 (2018)

Hansen, K.O. et al. 'Kinase chemodiversity from the Arctic: The breitfussins', *Journal of Medicinal Chemistry* 62: 10167–10181 (2019)

Kubota, S. 'Repeating rejuvenation in *Turritopsis*, an immortal hydrozoan (Cnidaria, Hydrozoa)', *Biogeography*: 101–103 (2011)

Martell, L. et al. 'Life cycle, morphology and medusa ontogenesis of *Turritopsis dohrnii* (Cnidaria: Hydrozoa)', *Italian Journal of Zoology* 83: 390–399 (2016)

Miglietta, M.P. et al. 'A silent invasion', *Biological Invasions* 11:825–834 (2009)

Piraino, S. et al. 'Reversing the life cycle: Medusae transforming into polyps and cell transdifferentiation in *Turritopsis nutricula* (Cnidaria, Hydrozoa)', *Biological Bulletin* 190: 302–312 (1996)

Tasdemir, D. 'Marine fungi in the spotlight: opportunities and challenges for marine fungal natural product discovery and biotechnology', *Fungal Biology and Biotechnology* 4: 5 (2017)

Wiegand, S. et al. 'Cultivation and functional characterization of 79 planctomycetes uncovers their unique biology', *Nature Microbiology* 5: 126–140 (2020)

Yoshinori, H. et al. 'De novo assembly of the transcriptome of *Turritopsis*, a jellyfish that repeatedly rejuvenates', *Zoological Science* 33: 366–371 (2016)

Securing the Foundations of Nature's Pharmacy

Europe's role: https://www.dw.com/en/europe-a-silent-hub-of-illegalwildlife-trade/a-37183459

Heinrich, S. et al. 'Where did all the pangolins go? International CITES trade in pangolin species', *Global Ecology and Conservation* 8: 241–253 (2016)

Lam, T.T.-Y. et al. 'Identifying SARS-CoV-2 related coronaviruses in Malayan pangolins', *Nature* (2020)

Mortality among reptiles trapped in the wild is so high it is comparable to that of cut flowers: https://www.jus.uio.no/ikrs/tjenester/kunnskap/kriminalpolitikk/meninger/2012/ulovlighandelmedtruededyrearter.html

Neergheen-Bhujun, V. et al. 'Biodiversity, drug discovery, and the future of global health: Introducing the biodiversity to biomedicine consortium, a call to action', *Journal of global health* 7: 020304–020304 (2017)

Operation Thunderbolt, June 2019: https://cites.org/eng/news/wildlife-trafficking-organized-crime-hit-hard-by-joint-interpol-wcoglobal-enforcement-operation_10072019

Pangolin removed from the list: https://www.nhm.ac.uk/discover/news/2020/june/china-removes-pangolin-scale-from-list-of-officialmedicines.html

Pimm, S.L. et al. 'The future of biodiversity', *Science* 269: 347 (1995)

Tigers in the US: https://www.theguardian.com/environment/shortcuts/2018/jun/20/more-tigers-live-in-us-back-yards-than-in-the-wild-is-this-a-catastrophe

5. The Fibre Factory

From Fluffy Seed to Favourite Fabric

Bank notes, Norway: https://www.norges-bank.no/tema/Sedler-og-mynter/

Bank notes, UK: https://www.bankofengland.co.uk/banknotes/currentbanknotes

Coppa, A. et al. 'Palaeontology: Early neolithic tradition of dentistry', *Nature* 440: 755–756 (2006)

FAO. *Measuring Sustainability in Cotton Farming Systems. Towards a Guidance Framework* (2015)

Mekonnen, M.M. et al. 'The green, blue and grey water footprint of crops and derived crop products', *Hydrology and Earth System Sciences* 15: 1577–1600 (2011)

Moulherat, C. et al. 'First evidence of cotton at neolithic Mehrgarh, Pakistan: Analysis of mineralized fibres from a copper bead', *Journal of Archaeological Science* 29: 1393–1401 (2002)

www.norges-bank.no/tema/Sedler-og-mynter/Ny-seddelserie/Om-sedlene/

Fact Sheet on Pesticide Use in Cotton Production. The Expert Panel on Social, Environmental and Economic Performance of Cotton Production (SEEP) (2012)

Splitstoser, J.C. et al. 'Early pre-Hispanic use of indigo blue in Peru', *Science Advances* 2: e1501623 (2016)

Three-quarters of all cotton grown is genetically modified: https://royalsociety.org/topics-policy/projects/gm-plants/what-gmcrops-are-currently-being-grown-and-where/

Home Sweet Home

28 stave churches: https://www.stavkirke.info/

Ålesund, city fire https://www.byggogbevar.no/pusse-opp/byggeskikk/jugendbyen-%C3%A5lesund

Concrete and CO_2: https://www.chathamhouse.org/sites/default/files/publications/2018-06-13-making-concrete-change-cement-lehne-preston-final.pdf

Fretheim, S.E. 'Mesolithic dwellings: An empirical approach to past trends and present interpretations in Norway', Doctoral thesis at NTNU; 2017:282 (2017)

Japan, new law in 2010: https://www.loc.gov/law/foreign-news/article/japan-law-to-promote-more-use-of-natural-wood-materials-for-public-buildings/

Kostenki Museum: https://www.rbth.com/history/329215-homosapiens-stone-age-russia and https://www.national

geographic.com/news/2014/11/141106-european-dna-
fossil-kostenki-science/

Que, Z.-L. et al. 'Traditional wooden buildings in China', *Wood in
Civil Engineering*. InTech (2017)

Seguin-Orlando, A. et al. 'Genomic structure in Europeans dating
back at least 36,200 years', *Science* 346: 1113 (2014)

By the Light of a Fungus Lamp

Desjardin, D.E. et al. 'Fungi bioluminescence revisited',
Photochemical & Photobiological Sciences 7: 170–182 (2008)

Purtov, K.V. et al. 'Why does the bioluminescent fungus *Armillaria
mellea* have luminous mycelium but non-luminous fruiting
body?' *Doklady Biochemistry and biophysics* 474:217–219
(2017)

Ramsbottom, J. *Mushrooms and Toadstools. A Study of the Activities
of Fungi.* Bloomsbury Books, 1953 (The quote from the war
correspondent comes from here.)

Sivinski, J. 'Arthropods attracted to luminous fungi', *Psyche* 88
(1981)

Chanterelle's Clever Cousins

Elven, H. et al. *Kunnskapsstatus for artsmangfoldet i Norge 2015*.
Utredning for Artsdatabanken 1/2016 (2016)

Guest, T. et al. 'Anticancer laccases: A review', *Journal of Clinical &
Experimental Oncology* 05 (2016)

Hakala, T.K. et al. 'Evaluation of novel wood-rotting polypores
and corticioid fungi for the decay and bio-pulping of Norway
spruce (*Picea abies*) wood', *Enzyme and Microbial Technology* 34:
255–263 (2004)

Patent application *Obba rivulosa*:
https://patents.google.com/patent/WO2003080812A1/en

Rashid, S. et al. 2011. 'A study of anti-cancer effects of *Funalia
trogii* in vitro and in vivo.' *Food and Chemical Toxicology* 49:
1477–1483 (2011)

Ibid. 'Potential of a *Funalia trogii* laccase enzyme as an anticancer agent', *Annals of Microbiology* 65 (2014)

Campfire Contemplation
A good third came from firewood, the rest from pellets, wood chips and liquid biofuel: https://nibio.no/tema/skog/bruk-av-tre/bioenergi

Sprucing Things Up: The Conifer that Flavours Food and Feeds Salmon
Ciriminna, R. et al. 'Vanillin: The case for greener production driven by sustainability megatrend', *Chemistry Open* 8: 660–667 (2019)

Crowther, T.W. et al. 'Mapping tree density at a global scale', *Nature* 525: 201–205 (2015)

Gallage, N.J. et al. 'Vanillin-bioconversion and bioengineering of the most popular plant flavor and its de novo biosynthesis in the vanilla orchid', *Molecular Plant* 8: 40–57 (2015)

Orchid vanilla price per kilo exceeds that of silver: https://www.foodbusinessnews.net/articles/13570-vanilla-prices-slowly-drop-as-crop-quality-improves

Øverland, M. et al. 'Yeast derived from lignocellulosic biomass as a sustainable feed resource for use in aquaculture', *Journal of the Science of Food and Agriculture* 97: 733–742 (2017)

Sahlmann, C. et al. 'Yeast as a protein source during smoltification of Atlantic salmon (*Salmo salar L.*), enhances performance and modulates health', *Aquaculture* 513: 734396 (2019)

Vanilla seeds can be added to ice cream for purely visual effect: https://www.cooksvanilla.com/vanilla-bean-seeds-a-troubling-new-trend/

6. The Caretaking Company

Too Much, Too Fast, Too Polluted

Berland, A. et al. 'The role of trees in urban stormwater management', *Landscape and urban planning* 162: 167–177 (2017)

Frazer, L. 'Paving paradise: The peril of impervious surfaces', *Environmental health perspectives* 113: A456–A462 (2005)

Grazing sheep on the roofs of Bergen: https://commons.wikimedia.org/wiki/Category:Hieronymus_Scholeus#/media/File:Scoleus.jpg

https://www.epa.gov/sites/production/files/2015-11/documents/stormwater2streettrees.pdf

https://extension.psu.edu/the-role-of-trees-and-forests-in-healthy-watersheds

https://www.sciencedaily.com/releases/2004/06/040615080052.htm

Magnussen, K. et al. '*Økosystemtjenester fra grønnstruktur i norske byer og tettsteder*', *Vista analyse* (2015)

When Money Grows on Trees

Bastin, J.-F. et al. 'Understanding climate change from a global analysis of city analogues', *PLOS ONE* 14: e0217592 (2019)

Huang, Y.J. et al. 'The potential of vegetation in reducing summer cooling loads in residential buildings', *Journal of Climate and Applied Meteorology* 26: 1103–1116 (1987)

IPBES. *The Global Assessment Report on Biodiversity and Ecosystem Services*. Complete draft version (2019)

London's most expensive tree: https://www.dailymail.co.uk/news/article-7733587/The-1-6million-tree-churchs-magnificent-marvel-valuable-specimen-UK.html#:~:text=After%20the%20system%20was%20launched,was%20valued%20at%20%C2%A3750%2C000.

Magnussen, K. et al. '*Økosystemtjenester fra grønnstruktur I norske byer og tettsteder*', *Vista analyse* (2015)

Nowak, D.J. et al. 'Air pollution removal by urban trees and shrubs in the United States', *Urban Forestry & Urban Greening* 4:115–123 (2006)

Trees reduce the temperature in cities: https://www.energylivenews.com/2019/09/30/could-urban-trees-mean-we-can-leave-airconditioning-emissions-behind/ and https://www.epa.gov/heatislands/using-trees-and-vegetation-reduce-heat-islands

Treeconomics London. Valuing London's Urban Forest Results of the London i-Tree Eco Project (2015)

Venter, Z.S. et al. 'COVID-19 lockdowns cause global air pollution declines with implications for public health risk', medRxiv: 2020.04.10.20060673 (2020a)

Ibid. 'Linking green infrastructure to urban heat and human health risk mitigation in Oslo, Norway', *Science of the Total Environment* 709: 136193 (2020b)

Wang, H. et al. 'Efficient removal of ultrafine particles from diesel exhaust by selected tree species: Implications for roadside planting for improving the quality of urban air', *Environmental Science & Technology* 53: 6906–6916 (2019)

How Green Was My Valley – Until the Topsoil Blew Away

'Fair is the slope ...' quotation from *Njål's Saga*, Wordsworth Classic edition (1998), translated by Carl F. Bayerschmidt and Lee M. Hollander

Sand lupines an extremely high ecological risk on the alien species list. 2018 https://artsdatabanken.no/Fab2018/N/1491

Flying Rivers in the Amazon

Diniz, M.B. et al. 'Does Amazonian land use display market failure? An opportunity-cost approach to the analysis of Amazonian environmental services', *CEPAL Review* 126: 99–118 (2018)

Ellison, D. et al. 'On the forest cover–water yield debate: from demand – to supply – side thinking', *Global Change Biology* 18:806–820 (2012)

Lindholm, M. '*Reguleres vinden av en biotisk pumpe?*' *Naturen* 138: 144–150 (2014)

Makarieva, A.M. et al. 'Biotic pump of atmospheric moisture as driver of the hydrological cycle on land', *Hydrology and EarthSystem Sciences* 11: 1013–1033 (2007)

Ibid. 'The biotic pump: Condensation, atmospheric dynamics and climate', *International Journal of Water* 5: 365–385 (2010)

Ibid. 'Where do winds come from? A new theory on how water vapor condensation influences atmospheric pressure and dynamics', *Atmospheric Chemistry and Physics* 13:1039–1056 (2013)

Sheil, D. et al. 'How forests attract rain: An examination of a new hypothesis', *BioScience* 59: 341–347 (2009)

Ibid. 'Forests, atmospheric water and an uncertain future: The new biology of the global water cycle', *Forest Ecosystems* 5 (2018)

Spracklen, D.V. et al. 'Observations of increased tropical rainfall preceded by air passage over forests', *Nature* 489: 282–285 (2012)

Ibid. 'Erratum: Corrigendum: Observations of increased tropical rainfall preceded by air passage over forests', *Nature* 494: 390–390 (2013)

Termites and Drought

Termites in the USA, damage: https://www.fs.fed.us/research/invasivespecies/insects/termites.php

Mangroves as Breakwaters

Arkema, K.K. et al. 'Linking social, ecological, and physical science to advance natural and nature-based protection for coastal communities', *Annals of the New York Academy of Science* 1399: 5–26 (2017)

Barbier, E.B. 'Valuing ecosystem services as productive inputs', *Economic Policy* 22: 177–229 (2007)

Ibid. 'The value of estuarine and coastal ecosystem services', *Ecological Monographs* 81: 169–193 (2011)

Das, S. et al. 'Mangroves protected villages and reduced death toll during Indian super cyclone', *PNAS* 106: 7357–7360 (2009)

Ibid. 'Mangroves can provide protection against wind damage during storms', *Estuarine Coastal and Shelf Science* 134:98 (2013)

Kathiresan, K. et al. 'Coastal mangrove forests mitigated tsunami', *Estuarine, Coastal and Shelf Science* 65: 601–606 (2005)

Russi, D., et al. 'The economics of ecosystems and biodiversity for water and wetlands', *TEEB Report* (2013)

Thomas, N., et al. 'Distribution and drivers of global mangrove forest change, 1996–2010', *PLOS ONE* 12: e0179302 (2017)

Beauty in a Rotting Branch

Excerpt from Tarjei Vesaas' 'Trøytt tre' (one of my favourite poems, is from the collection), *Lykka for ferdesmenn* (1949)

Jacobsen, R.M. et al. 'Near-natural forests harbor richer saproxylic beetle communities than those in intensively managed forests', *Forest Ecology and Management* 466: 118124 (2020)

Norden, J. et al. 'At which spatial and temporal scales can fungi indicate habitat connectivity?' *Ecological Indicators* 91: 138–148 (2018)

Pennanen, J. 'Forest age distribution under mixed-severity fire regimes – a simulation-based analysis for middle boreal Fennoscandia', *Silva Fennica* 36: 213–231 (2002)

Of Reindeer and Ravens

Badia, R. 'Reindeer carcasses provide foraging habitat for several bird species of the alpine tundra', *Ornis Norvegica* 42: 36–40 (2019)

Carter, D.O. et al. 'Cadaver decomposition in terrestrial ecosystems', *Naturwissenschaften* 94: 12–24 (2006)

Frank, S.C. et al. 'Fear the reaper: Ungulate carcasses may generate an ephemeral landscape of fear for rodents', *Royal Society Open Science* 7: 191644 (2020)

Granum, H.M. 'Change in arthropod communities following a mass death incident of reindeer at Hardangervidda', Master thesis NMBU (2019)

Payne, J.A. 'A summer carrion study of the baby pig *Sus Scrofa Linnaeus*', *Ecology* 46: 592–602 (1965)

Personally communicated by Sam Steyaert.

Steyaert, S.M.J.G. et al. 'Special delivery: Scavengers direct seed dispersal towards ungulate carcasses', *Biology Letters* 14:20180388 (2018)

The poem 'The Raven' by Edgar Allan Poe can be read here: https://www.poetryfoundation.org/poems/48860/the-raven

7. The Warp in the Tapestry of Life

Chisholm, S.W. et al. 'A novel free-living prochlorophyte abundant in the oceanic euphotic zone', *Nature* 334: 340–343 (1988)

Flombaum, P. et al. 'Present and future global distributions of the marine cyanobacteria *Prochlorococcus* and *Synechococcus*', *PNAS* 110: 9824–9829 (2013)

The scientist who dedicated her career to *Prochlorococcus* is Penny Chisholm – you can read about her here: https://www.sciencemag.org/news/2017/03/meet-obscure-microbe-influences-climate-ocean-ecosystems-and-perhaps-even-evolution

Whale Fall and White Gold

Cushman, G.T. *Guano and the Opening of the Pacific World. A Global Ecological History*, Cambridge University Press, 2013

Danovaro, R. et al. 'The deep-sea under global change', *Current Biology* 27: R461–R465 (2017)

Doughty, C.E. et al. 'Global nutrient transport in a world of giants', *PNAS* 113: 868–873 (2016)

Glover, A.G. et al. 'World-wide whale worms? A new species of Osedax from the shallow north Atlantic', *Proceedings of the Royal Soc. B*: 272:2587–2592 (2005)

Kjeld, M. 'Salt and water balance of modern baleen whales: Rate of urine production and food intake', *Canadian Journal of Zoology* 81: 606–616 (2003)

LaRue, M.A. et al. 'Emigration in emperor penguins: implications for interpretation of long-term studies', *Ecography* 38: 114–120 (2015)

Otero, X.L. et al. 'Seabird colonies as important global drivers in the nitrogen and phosphorus cycles', *Nature Communications* 9 (2018)

Roman, J. et al. 'Whales as marine ecosystem engineers', *Frontiers in Ecology and the Environment* 12: 377–385 (2014)

Rouse, G.W. 'Osedax: Bone-eating marine worms with dwarf males', *Science* 305: 668–671 (2004)

The Guano Island Act: https://uscode.house.gov/view.xhtml?path=/prelim@title48/chapter8&edition=prelim

The World's Most Beautiful Carbon Store

Achat, D.L. et al. 'Forest soil carbon is threatened by intensive biomass harvesting', *Scientific Reports* 5: 15991 (2015)

Bartlett, J. et al. *Carbon storage in Norwegian ecosystems*, NINA Report1774, Norwegian Institute for Nature Research (2020)

IPCC. *Climate Change 2014: Synthesis Report. Contribution of Working Groups I, II and III to the Fifth Assessment Report of the Intergovernmental Panel on Climate Change*, IPCC, Geneva, Switzerland (2014)

Kroeker, K.J. et al. 'Meta-analysis reveals negative yet variable effects of ocean acidification on marine organism', *Ecology Letters* 13: 1419–1434 (2010)

Luyssaert, S. et al. 'Old-growth forests as global carbon sinks', *Nature* 455: 213–215 (2008)

Healthy Nature Regulates Disease

American Veterinary Medical Association. 'One Health: A New Professional Imperative', One Health Initiative Task Force: Final Report (2008)

Blockstein, D.E. et al. 'Fauna in decline: Extinct pigeon's tale', *Science* 345: 1129 (2014)

Jørgensen, H.J. et al. 'COVID-19: Én verden,én helse'. *Tidsskrift for Den norske legeforening* 140. doi:10.4045/tidsskr.20.0212 (2020)

Keesing, F. et al. 'Hosts as ecological traps for the vector of Lyme disease', *Proceedings of the Royal Society B: Biological Sciences* 276: 3911–3919 (2009)

Link with Lyme disease: https://www.wiscontext.org/what-does-passenger-pigeon-have-do-lyme-disease

Nesting place in Michigan: https://sora.unm.edu/sites/default/files/journals/nab/v039n05/p00845-p00851.pdf

Ostfeld, R.S. et al. 'Effects of acorn production and mouse abundance on abundance and *Borrelia burgdorferi* infection prevalence of nymphal *Ixodes scapularis* ticks', *Vector-Borne and Zoonotic Diseases* 1: 55–63 (2001)

Ibid. 'Are predators good for your health? Evaluating evidence for top-down regulation of zoonotic disease reservoirs', *Frontiers in Ecology and the Environment* 2: 13–20 (2004)

Ibid. 'Tick-borne disease risk in a forest food web'. *Ecology* 99: 1562–1573 (2018)

Passenger pigeon numbers in the past: https://www.si.edu/spotlight/passengerpigeon

Rohr, J.R. et al. 'Emerging human infectious diseases and the links to global food production', *Nature Sustainability* 2: 445–456 (2019)

Settele, J. et al. 'COVID-19 stimulus measures must save lives, protect livelihoods, and safeguard nature to reduce the risk of future pandemics', IPBES Expert Guest Article (2020)

Tanner, E. et al. 'Wolves contribute to disease control in a multihost system', *Scientific Reports* 9 (2019)

Taylor, L.H. et al. 'Risk factors for human disease emergence', *Philosophical Transactions of the Royal Society London B: Biological Sciences* 356: 983–989 (2001)

World Health Organization and Convention on Biological Diversity. *Connecting Global Priorities: Biodiversity and Human Health. A State of Knowledge Review*. 364 p. WHO, Geneva (2015)

The Very Hungry Caterpillar

Bohan, D.A. et al. 'National-scale regulation of the weed seedbank by carabid predators', *Journal of Applied Ecology* 48: 888–898 (2011)

Hass, A.L. et al. 'Landscape configurational heterogeneity by small-scale agriculture, not crop diversity, maintains pollinators and plant reproduction in Western Europe', *Proceedings of the Royal Society B: Biological Sciences* 285: 20172242 (2018)

Lechenet, M. et al. 'Reducing pesticide use while preserving crop productivity and profitability on arable farms', *Nature Plants* 3: 17008 (2017)

Roslin, T. et al. 'Higher predation risk for insect prey at low latitudes and elevations', *Science* 356: 742–744 (2017)

Tscharntke, T. et al. 'Multifunctional shade-tree management in tropical agroforestry landscapes – a review', *Journal of Applied Ecology* 48: 619–629 (2011)

Tschumi, M. et al. 'High effectiveness of tailored flower strips in reducing pests and crop plant damage', *Proceedings of the Royal Society B: Biological Sciences* 282: 20151369 (2015)

United Nations. *Report of the Special Rapporteur on the Right to Food (A/HRC/34/48). UN Human Rights Council.* https://documents-dds-ny.un.org/doc/UNDOC/GEN/G17/017/85/PDF/G1701785.pdf?OpenElement= (2017)

Wetherbee, Birkemoe, Sverdrup-Thygeson, in prep. 'Veteran trees as a source of natural enemies'.

8. Nature's Archives

Ray Bradbury quotation: 'Without libraries, what have we? We have no past and no future' from https://www.goodreads.com/quotes/145695-without-libraries-what-have-we-we-haveno-past-and

When Pollen Speaks
Bryant, V.M. et al. 'Forensic palynology: A new way to catch crooks', Bryant, V.M. and Wrenn, J.W. (eds.), *New Development in Palynomorph Sampling, Extraction, and Analysis; American Association of Stratigraphic Palynologists Foundation, Contributions Series* Number 33, 145–155 (1998)

Holloway, R. et al. 'New directions in palynology in ethnobiology', *Journal of Ethnobiology* 6: 46–65 (1986)

Milne, L. et al. 'Forensic palynology', I.H.M. Coyle (ed.), *Forensic Botany: Principles and Applications to Criminal Casework* (pp. 217–252). Boca Raton, USA: CRC Press, 2005

Sandom, C.J. et al. 'High herbivore density associated with vegetation diversity in interglacial ecosystems', *PNAS* 111:4162–4167 (2014)

Smith, D. et al. 'Can we characterise "openness" in the Holocene palaeoenvironmental record? Modern analogue studies of insect faunas and pollen spectra from Dunham Massey deer park and Epping Forest, England', *The Holocene* 20: 215–229 (2010)

Steele, A. et al. 'Reconstructing Earth's past climates: The hidden secrets of pollen', *Science Activities: Classroom Projects and Curriculum Ideas* 50: 62–71 (2013)

Stephen, A. 'Pollen – A microscopic wonder of plant kingdom', *International Journal of Advanced Research in Biological Sciences*, 1: 45–62 (2014)

The verse is the opening of 'Auguries of Innocence' by William Blake, first published in 1863

Whitehouse, N.J. et al. 'How fragmented was the British Holocene wildwood? Perspectives on the "Vera" grazing debate from the fossil beetle record', *Quaternary Science Reviews* 29:539–553 (2010)

Rings of Lived Life

Bill, J. et al. 'The plundering of the ship graves from Oseberg and Gokstad: an example of power politics?' *Antiquity* 86: 808–824 (2012)

Buntgen, U. et al. '2500 years of European climate variability and human susceptibility', *Science* 331: 578–582 (2011)

Grissino-Mayer, H.D. et al. 'Tree-ring dating of the Karr-Koussevitzky double bass: A case study' in *Dendromusicology* 61: 77–86 (2005)

Rolstad, J. et al. 'Fire history in a western Fennoscandian boreal forest as influenced by human land use and climate', *Ecological Monographs* 87: 219–245 (2017)

The cited verse is from Hans Børli's poem '*Fra en tømmerhoggers dagbok*', published in the collection *Dag og drøm. Dikt i utvalg*. H. Aschehoug & Co., 1979

Chimney Talks Crap

BirdLife International. *Chaetura pelagica*, The IUCN Red List of Threatened Species 2018: e.T22686709A131792415. https://dx.doi.org/10.2305/IUCN.UK.2018-2.RLTS. T22686709A131792415.en (2018)

English, P.A. et al. 'Stable isotopes from museum specimens may provide evidence of long-term change in the trophic ecology of a migratory aerial insectivore', *Frontiers in Ecology and Evolution* 6:1–13 (2018)

Nocera, J.J. et al. 'Historical pesticide applications coincided with an altered diet of aerially foraging insectivorous chimney swifts', *Proceedings of the Royal Society B: Biological Sciences* 279: 3114–3120 (2012)

9. An Ideas Bank for Every Occasion

Sacred Lotus with Self-cleaning Surfaces
Barthlott, W. et al. 'Purity of the sacred lotus, or escape from contamination in biological surfaces', *Planta* 202: 1–8 (1997)

Shen-Miller, J. et al. 'Exceptional seed longevity and robust growth: Ancient sacred lotus from China', *American Journal of Botany* 82: 1367–1380 (1995)

Shirtcliffe, N.J. et al. 'Learning from superhydrophobic plants: The use of hydrophilic areas on superhydrophobic surfaces for droplet control'. Part of the *Langmuir 25th Year: Wetting and Superhydrophobicity* special issue 25: 14121–14128 (2009)

The haiku by Matsuo Basho is from the translation *Narrow Road to the Interior: And Other Writings*, translated by Sam Hamill, Shambhala Publications, 2006

Zygote Quarterly (digital magazine about bioinspiration) 3, 2012: https://zqjournal.org/editions/zq03.html

Shinkansen – A Bird-beaked Bullet Train
About the Shinkansen:
https://www.greenbiz.com/blog/2012/10/19/how-one-engineers-birdwatching-made-japans-bullet-train-better
https://asknature.org/resource/the-world-is-poorly-designed-but-copying-nature-helps/
https://www.aftenposten.no/verden/i/xP5l8j/naa-skal-de-testkjoerelyntog-med-toppfart-paa-400-kmt

Rao, C. et al. 'Owl-inspired leading-edge serrations play a crucial role in aerodynamic force production and sound suppression', *Bioinspiration & Biomimetics* 12: 046008 (2017)

Wagner, H., et al. 'Features of owl wings that promote silent flight', *Interface* Focus 7: 20160078 (2017)

Colours That Never Fade

Fayemi, P.-E. et al. 'Bio-inspired design characterisation and its links with problem-solving tools', *Design* 2014 Dubrovnik – Croatia (2014)

L'Oréal: http://canadianbeauty.com/luci-from-lancome/ and https://www.temptalia.com/lancome-spring-2008-luci-luminous-colorless-color-intelligence-collection/

Shu, L.H. et al. 'Biologically inspired design', *CIRP Annals* 60: 673–693 (2011)

Sun, J. et al. 'Structural coloration in nature', *RSC Adv.* 3: 14862–14889 (2013)

Vukusic, P. 'An introduction to bio-inspired design. Nature's inspiration may help scientists find solutions to technological, biomedical or industrial challenges', *Contact Lens Spectrum* (2010)

Zhang, S. et al. 'Nanofabrication and coloration study of artificial *Morpho* butterfly wings with aligned lamellae layers', *Scientific Reports* 5: 16637 (2015)

Moths with an Eye for Darkness

Bixler, G.D. et al. 'Biofouling: Lessons from nature', *Philosophical Transactions of the Royal Society A: Mathematical, Physical and Engineering Sciences* 370: 2381–2417 (2012)

Examples of bioinspired products mentioned:

https://www.geomatec.com/products-and-solutions/optical-control/anti-reflection-and-anti-glare/gmoth/

https://www.m-chemical.co.jp/en/products/departments/mcc/hpfilms-pl/product/1201267_7568.html

https://www.sharklet.com/

https://web-japan.org/trends/11_tech-life/tec201901.html

Hirose, E. et al. 'Does a nano-scale nipple array (moth-eye structure) suppress the settlement of ascidian larvae?', *Journal of the Marine Biological Association of the United Kingdom*, 99:1393–1397 (2019)

Navarro-Baena, I. et al. 'Single-imprint moth-eye anti-reflective and self-cleaning film with enhanced resistance', *Nanoscale* 10:15496–15504 (2018)

Sun, J. et al. 'Biomimetic moth-eye nanofabrication: Enhanced antireflection with superior self-cleaning characteristic', *Scientific Reports* 8: 1–10 (2018)

Tan, G. et al. 'Broadband antireflection film with moth-eye-like structure for flexible display applications', *Optica* 4: 678 (2017)

Smart as Slime Mould

Adamatzky, A. et al. 'Are motorways rational from slime mould's point of view?', *International Journal of Parallel, Emergent and Distributed Systems*, 28: 230–248 (2013)

Navlakha, S. et al. 'Algorithms in nature: The convergence of systems biology and computational thinking', *Molecular Systems Biology*, 7: 546 (2011)

Poissonnier, L.-A. et al. 'Experimental investigation of ant traffic under crowded conditions', *eLife* 8 (2019)

Slime mould mating types: https://warwick.ac.uk/fac/sci/lifesci/outreach/slimemold/facts/

Sørensen, K. 'Metaheuristics – the metaphor exposed', *International Transactions in Operational Research*, 22: 3–18 (2015)

Tero, A. et al. 'Rules for biologically inspired adaptive network design', *Science* 327: 439–442 (2010)

The honeybee algorithm: https://www.goldengooseaward.org/awardees/honey-bee-algorithm

The Hermit Beetle and the Hound

Middle, I. 'Between a dog and a green space: Applying ecosystem services theory to explore the human benefits of off-the-leash dog parks', *Landscape Research*: 1–14 (2019)

Mosconi, F. et al. 'Training of a dog for the monitoring of *Osmoderma eremita*', *Nature Conservation-Bulgaria*: 237–264 (2017)

Other examples of use of dogs:
https://www.greenmatters.com/news/2018/02/19/2m3wBf/
   bordercollies-forest
https://www.iowapublicradio.org/post/specially-trained-dogs-help-
   conservationists-find-rare-iowa-turtles
https://www.bbcearth.com/blog/?article=meet-the-dogs-
   savingendangered-species
Sverdrup-Thygeson, A. et al. *Faglig grunnlag for handlingsplan for
   eremitt*, NINA Rapport 632. 25 pages (2010)
The excerpt is from Rudyard Kipling's 'The Cat that Walked by
   Himself', originally published in *Just So Stories*, Macmillan
   Publishers, 1902

Like a Bat into Hell

Christen, A.G. et al. 'Dr. Lytle Adams' incendiary "bat bomb" of
   World War II', *Journal of the History of Dentistry*, 52: 109–115
   (2004)
Excerpt from the Tarjei Vesaas' poem '*Regn i Hiroshima*' (Rain in
   Hiroshima) from *Leiken og Lynet* (1947); English translation
   from *Through Naked Branches: Selected Poems of Tarjei Vesaas*,
   translated by Roger Greenwald, Black Widow Press, 2018
Pigeon-controlled missiles: https://web.archive.org/web/
   20080516215806/http://historywired.si.edu/object.cfm?ID=353
The Dickin Medal: https://www.pdsa.org.uk/what-we-do/
   animal-awards-programme/pdsa-dickin-medal

10. Nature's Cathedral – Where Great
Thoughts Take Shape

The excerpt from *Voluspå* is taken from Penguin Classics edition
   of *The Prose Edda* translated by Jesse Byock, 2005
Ekman, Kerstin. *My Life in the Forest and the Forest in my Life –
   Nature and Identity, Herrene i skogen*, Heinesen Forlag. 2015

Homo indoorus – *Nature and Health*

Bratman, G.N. et al. 'Nature and mental health: An ecosystem service perspective', *Science Advances* 5: eaax0903 (2019)

Chawla, L. 'Childhood experiences associated with care for the natural world: A theoretical framework for empirical results', *Children, Youth and Environments*, 17: 144–170 (2007)

Example from the UK: https://www.dailymail.co.uk/news/article-462091/How-children-lost-right-roam-generations.html

Haahtela, T. et al. 'The biodiversity hypothesis and allergic disease: World allergy organization position statement', *World Allergy Organization Journal*, 6: 3 (2013)

Mayer, F.S. et al. 'The connectedness to nature scale: A measure of individuals' feeling in community with nature', *Journal of Environmental Psychology*, 24: 503–515 (2004)

Mental health problems in Norwegians: https://www.helsenett.no/142-fakta/fakta.html

Nilsen, A.H. 'Available outdoor space and competing needs in public kindergartens in Oslo', *FORMakademisk* 7 (2014)

Ohtsuka, Y. et al. 'Shinrin-yoku (forest-air bathing and walking) effectively decreases blood glucose levels in diabetic patients', *International Journal of Biometeorology*, 41: 125–127 (1998)

Sandifer, P.A. et al. 'Exploring connections among nature, biodiversity, ecosystem services, and human health and wellbeing: Opportunities to enhance health and biodiversity conservation', *Ecosystem Services*, 12: 1–15 (2015)

Sender, R. et al. 'Revised estimates for the number of human and bacteria cells in the body', *PLOS Biology*, 14: e1002533 (2016)

Brink P. et al. *The Health and Social Benefits of Nature and Biodiversity Protection. A report for the European Commission*, Institute for European Environmental Policy, London/Brussels (2016)

Van Den Berg, A.E. 'From green space to green prescriptions: Challenges and opportunities for research and practice', *Frontiers in Psychology* 8 (2017)

Von Hertzen, L. et al. 'Natural immunity', *EMBO reports* 12:1089–1093 (2011)

Wells, N. et al. 'Nature and the life course: Pathways from childhood nature experiences to adult environmentalism', *Children, Youth and Environments* 16 (2006)

World Health Organization and Convention on Biological Diversity. 2015. *Connecting Global Priorities: Biodiversity and Human Health. A State of Knowledge Review*. 364 p. WHO, Geneva

Smart as a Plant – Other Species Can Do More Than You Think

Appel, H.M. et al. 'Plants respond to leaf vibrations caused by insect herbivore chewing', *Oecologia* 175: 1257–1266 (2014)

Balding, M. et al. 'Plant blindness and the implications for plant conservation', *Conservation Biology* 30: 1192–1199 (2016)

Biegler, R. 'Insufficient evidence for habituation in *Mimosa pudica*. Response to Gagliano et al', (2014). *Oecologia* 186: 33–35 (2018)

Gagliano, M. et al. 'Experience teaches plants to learn faster and forget slower in environments where it matters', *Oecologia* 175:63–72 (2014)

Ibid. 'Learning by Association in Plants', *Scientific Reports* 6: 38427 (2016)

Ibid. 'Plants learn and remember: let's get used to it', *Oecologia* 186: 29–31 (2018)

Helms, A.M. et al. 'Exposure of *Solidago altissima* plants to volatile emissions of an insect antagonist (*Eurosta solidaginis*) deters subsequent herbivory', *PNAS* 110: 199–204 (2013)

Knapp, S. 'Are humans really blind to plants?', *Plants, People, Planet* 1: 164–168 (2019)

Mescher, M.C. et al. 'Plant biology: Pass the ammunition', *Nature* 510: 221–222 (2014)

Pierik, R. et al. 'Molecular mechanisms of plant competition: Neighbour detection and response strategies', *Functional Ecology* 27: 841–853 (2013)

Runyon, J.B. et al. 'Volatile chemical cues guide host location and host selection by parasitic plants', *Science* 313: 1964 (2006)

Veits, M. et al. 'Flowers respond to pollinator sound within minutes by increasing nectar sugar concentration', *Ecology Letters* 22: 1483–1492 (2019)

With a Little Help from My Friends – An Intricate Interaction

Brusca, R. et al. 'Tongue replacement in a marine fish (*Lutjanus guttatus*) by a parasitic isopod (Crustacea: Isopoda)', *Copeia* 1983: 813 (1983)

Fay, M.F. 'Orchid conservation: How can we meet the challenges in the twenty-first century?', *Botanical Studies* 59 (2018)

NOU 2004: 28. Act relating to the management of natural, landscape and biological diversity (Nature Diversity Act). Ministry of Climate and Environment (2004)

Lost Wilderness and New Nature: The Way Forward

Perring, M.P. et al. 'The extent of novel ecosystems: Long in time and broad in space', In Hobbs, R. J. et al. (eds), *Novel Ecosystems*, pp. 66–80 (2013)

The Aldo Leopold quotation is from *Round River: From the Journals of Aldo Leopold*, Northwood Press, 1953

The Henry Thoreau quote is from *Walden or Life in the Woods*, originally published in 1854

Vizentin-Bugoni, J. et al. 'Structure, spatial dynamics, and stability of novel seed dispersal mutualistic networks in Hawaii', *Science* 364: 78–82 (2019)

Watson, J.E.M. et al. 'Catastrophic declines in wilderness areas undermine global environment targets', *Current Biology* 26: 2929–2934 (2016)

Ibid. 'Protect the last of the wild', *Nature* 563:27–30 (2018)

Afterword

IPBES. *The Global Assessment Report on Biodiversity and Ecosystem Services.* Complete draft version (2019)
Statistic from Our World in Data, https://ourworldindata.org/
The quotation at the beginning is by Arundhati Roy, author and activist, from a speech at the World Social Forum in Porto Alegre, Brazil, 2003

Index